T0252437

ENERGY WITHOUT CONSCIENCE

ENERGY WITHOUT CONSCIENCE

Oil, Climate Change, and Complicity

David McDermott Hughes

DUKE UNIVERSITY PRESS *Durham and London* 2017

© 2017 Duke University Press
All rights reserved
Printed and bound by CPI Group (UK) Ltd, Croydon, CR0 4YY
Cover designed by Matthew Tauch
Typeset in Arno Pro & Meta by Graphic Composition, Inc.,
Bogart, GA.

Library of Congress Cataloging-in-Publication Data
Names: Hughes, David McDermott, author.
Title: Energy without conscience : oil, climate change, and
complicity / David McDermott Hughes.
Description: Durham : Duke University Press, 2017. | Includes
bibliographical references and index.
Identifiers: LCCN 2016035965 (print) | LCCN 2016037765 (ebook)
ISBN 9780822363064 (hardcover : alk. paper)
ISBN 9780822362982 (pbk. : alk. paper)
ISBN 9780822373360 (e-book)
Subjects: LCSH: Energy industries—Environmental aspects. |
Energy industries—Moral and ethical aspects. | Slavery—
Trinidad and Tobago—Trinidad—History. | Petroleum industry
and trade—Colonies—Great Britain. | Petroleum industry and
trade—Trinidad and Tobago—Trinidad.
Classification: LCC HD9502.T72 H84 2017 (print) | LCC HD9502.T72
(ebook) | DDC 338.2/72820972983—dc23
LC record available at https://lccn.loc.gov/2016035965

Cover credit: *Close-up of pitch at the world's largest natural pitch lake,
Trinidad*, 2007. Photo © Robert Harding.

FOR JESSE AND SOPHIA

CONTENTS

ACKNOWLEDGMENTS

As a scholar of southern Africa, I came to Trinidad and Tobago rather unprepared. Individuals, rather than institutions, took me under their wing and inducted me into the secrets of the country and its oil sector. Among these friends, I want especially to thank Gerard and Alice Besson, Joan Dayal, Dax Driver, Simone Mangal, Jeremy and Michelle Matouk, Patricia Mohammed, Krishna Persad, Marina Salandy-Brown, Mary Schorse, Eden Shand, Teresa White, and Mark Wilson. My informants, who include many of these people, assisted the research and, obviously, made it possible. These women and men are too many to name, and more than a few wish their identities to remain confidential. In the course of the research and writing of this book, four of my informants passed away: Norris Deonarine, Rhea Mungal, Denis Pantin, and Julian Kenny. In all but the last case, these environmental activists died young and unexpectedly. Their loss impoverishes Trinidad and Tobago of voices that could grapple critically with hydrocarbons. Some of the protagonists in this book will disagree violently with its tone and conclusions. I hope they will find their views fairly represented, if also sufficiently refracted to teach something new.

In the United States, colleagues and student colleagues helped beat the manuscript into shape. With gratitude, I acknowledge Hannah Appel, Jacob Campbell, Isaac Curtis, Daniel Goldstein, Angelique Haugerud, Dorothy Hodgson, Judith Hughes, Enrique Jaramillo, Mazen Labban, Arthur Mason, Melanie McDermott, Benjamin Orlove, Peter Rudiak-Gould, Marian Thorpe, Michael Watts, and Paige West. For images, primary documents, critical commentary, or pivotal conversations, I am indebted to Andrew Matthews, Gerard Besson, Selwyn Cudjoe, Marlaina Martin, Mike Siegel, Genese Sodikoff, Steven Stoll, Humphrey Stollmeyer, Anna Tsing, and Richard York. I benefited from speaking engagements at Bard College, Brown University, Carleton University, Columbia University, Dartmouth

College, DePauw University, the International Institute of Social Studies, New York University, the New York Academy of Sciences, Rice University, University of Alberta, University of Leeds, and Wellesley College. The Rutgers Center for Cultural Analysis provided me with a small fellowship to assist in writing the manuscript. I acknowledge the journal *American Anthropologist* for allowing me to publish, as chapter 5, a revised version of the article "Climate Change and the Victim Slot: From Oil to Innocence" (115, no. 4 [2013]: 570–81). I also thank McGill-Queen's University Press for allowing me to include, in chapter 4, text from my contribution to the volume *Petrocultures: Oil, Energy, Culture* (2017), edited by Sheena Wilson, Adam Carlson, and Imre Szeman. Finally, Gisela Fosado at Duke University Press believed in this project before I knew what it was and brought the ship in to shore with skill and grace.

Beyond this one book, I benefit from an infrastructure that is both political and emotional. The climate justice movement and 350.org allow me to anticipate an energy transition that will happen. Without such hope, the manuscript would have slipped into a despair too tedious to read. The academic labor movement—and its faculty union at Rutgers—cultivates a different kind of hope: the affirmation that fieldwork, writing, teaching, and university service still constitute a life's calling. And in that life, there is no one more important than my wife, Melanie, who has supported and sacrificed for this calling over two decades.

It would seem grandiose to dedicate *Energy without Conscience* to the billions-strong victims of fossil fuels. So, with Melanie, I choose two people enduring and resisting climate change: our children, Jesse and Sophia. Beginning in Trinidad, you have expressed curiosity, concern, outrage, and activism for a sustainable and just world. May you inspire others to feel and do the same.

How does it feel to change the climate? This question seems more absurd than impolite. It implies a chain of causation and responsibility that still remains invisible and mostly unacknowledged. In fact, some people—a billion high emitters—burn oil and otherwise pump carbon dioxide (CO_2) into the atmosphere at a rate dangerous to societies and ecosystems everywhere (Chakravarty et al. 2010). A slice of this population—overrepresented in the United States—disputes the science and scenarios of climate change. But explicit denial is less widespread than silence and disregard. The bulk of informed consumers simply don't care a great deal about carbon emissions and their consequences. Tobacco provokes stronger reactions, indeed sometimes a disgust verging on revulsion. Where is the revulsion over flood, drought, and myriad other catastrophic shifts in the conditions for life and society on planet Earth? Menacing as it increasingly is, climate change has yet to become a moral issue for most people.

Energy without Conscience seeks to explain this persistent banality. I am not trying to expose—as others have done—the greed of individuals, firms, or governments. Capitalism and convenience certainly underwrite the status quo. Yet means-to-ends reasoning does not account fully for the abundance of support for fossil fuels. Cultural meanings also sustain hydrocarbons. In the oil profession itself, people drill for noneconomic, as well as economic, motives. "The romance [among oil geologists] was not really based on money, which was only a way of keeping score," reminisces the Texan John Graves (1995, xi–xii) in an essay on prospecting. His nostalgia exceeds his greed. I am interested in such cultural dispositions and discourses. As I argue, they obscure responsibility for carbon emissions among those most responsible and those most susceptible—technicians in and local bystanders to the fossil fuel business (who are often the same people). Certain modes of thought inside and outside the industry push a more critical consideration of oil to the margin. Hydrocarbons—as I refer to oil, natural gas, coal, and bitumen—seem both invisible and inevitable. One notices them only when something goes wrong—when, for instance, massive volumes gush into the Gulf of Mexico. Water-borne pollution of this sort triggers professional concern as well as public outrage. This book, on the other hand, describes the everyday, intended functions of our energy system. When platforms, pipelines, and pumps work properly, oil arrives safely at the gas tank of a motor vehicle. Then, combusted in the engine, the hydrocarbon spews carbon dioxide into the air unnoticed and without protest. One might refer to this form of pollution as "the spill everywhere." It far outweighs local contamination, both in volume and in planetary effects. Oil, in other words, is most dangerous when it behaves ordinarily and when people treat it as ordinary—that is, as neither moral nor immoral, but amoral.

Investigating such a nonevent—really the partial absence of meaning—requires an indirect approach. One has to detect the meaning and sentiment that prevent an accumulation of feeling around oil or carbon emissions. Why do hydrocarbons not inspire disgust—or romance for that matter—among more people more often? To answer this question, one has to measure the subtle effort expended as informed people avoid reflecting ethically or emotionally upon oil. The right circumstances will throw this making of ordinariness into the sharpest relief. I found those conditions at the birthplace of petroleum: Trinidad in the southern Caribbean (map 1.1). Here, Walter Darwent drilled the world's first continually productive oil well in 1866.[1] This larger island of Trinidad and Tobago shares deposits with nearby Venezuela. Until recently, it contributed the lion's share of gas imported to the United States. But it does not rank among the traditional petrostates, either in production or in reputation. I lived in Port of Spain, the capital of Trinidad and Tobago, for the 2009–10 academic year and conducted ethnographic research among energy experts, anti-industrial activists, and policy makers preparing for climate change. At that point, Trinidad (as I abbreviate the nation-state) had never suffered a major spill. In terms of environmental harm, the industry was primarily committing climate change through CO_2 emissions. But Trinidadians—whose per capita carbon emissions ranked fourth among nations—did not appreciate this responsibility. My informants considered themselves to be victims—and only victims—of rising seas. In these ways, groups of

MAP I.1 Trinidad. Prepared by Mike Siegel of Rutgers Cartography Lab.

Trinis edged so close to the moral problem of hydrocarbons that they had to avert their gaze. Looking historically at Trinidad's energy systems, as I do in part I, I found moments when energy both did and did not prick the conscience. Plantation slavery—reliant upon embodied, somatic power—never achieved stability. Bonded people constantly reminded masters and governors of the bondsmen's individuality, of their will for freedom. Conscience dogged the energy that harvested sugar. Hydrocarbons arrived with no such baggage. Petroleum raised no moral outrage or endorsement, and contemporary beliefs, institutions, and forms of expertise helped to keep it that way. (Coal, a notable absence, has never been produced in Trinidad.) That process of overlooking consequences continues today. *Energy without Conscience* illuminates the people close to and conducting this work—subjects both intimate with and untroubled by the carbon bomb ticking around them.

I did not approach these women and men dispassionately, and I have not written about them with the usual ethnographic sympathy. Frankly, I oppose their interests. Partiality is not new to my field: anthropologists often take sides, engaging with popular movements and local projects (Goldstein 2012, 35ff.). Nancy Scheper-Hughes advocates a "militant anthropology," eschewing "false neutrality . . . in the face of the broad political dramas of life and death, good and evil" (1995, 411). In solidarity, she joined desperately poor mothers of a Brazilian shantytown as a *companheira*. Stop merely spectating, she demands of anthropologists. Practice instead an "ethic of care and responsibility" toward your informants (419). I have answered that call only halfway. From the beginning, I encountered oil as immoral—and as an industry that should go extinct. I hope for a rapid and complete conversion to wind and solar power, a change both necessary and, experts increasingly suggest, feasible as well (Jacobson and Delucchi 2009). We may still need oil for plastics and for some kinds of high-reliability energy uses, in hospitals, for example. Undeniably, however, I wish an end to the current livelihoods of most of the people—even of my friends—described in this book. Therefore, I do not express care toward petroleum geologists. I write about them with understanding and with ethnographic nuance, but I shall not present myself as a companheiro in relation to this social group. Besides, my subjects never asked for care, comradeship, or solidarity. Wealthy and powerful, they need no help from scholars. Hence, a militant anthropology of elites can afford a certain tension, emphasizing responsibility more than care. There is a difference between these two attitudes. The responsible writer looks over an informant's shoulder, prepared to reveal and criticize the wider harm that person may cause. Perhaps this is where the social science of climate change needs to go: resisting fossil fuels by documenting how their promoters think, act, and feel. Complicity, in a word, is the chief concern of this book.

The Ethical Deficit

I arrived in Trinidad expecting abundant art and literature about oil and gas. Those two commodities, after all, drove the leading industry in this acknowledged petrostate. I thought I knew how to trace the links between energy systems and cultural expression. At that very moment, I was in the process of publishing my second book on Zimbabwe (Hughes 2010). The ethnography concerned white Zimbabweans, including their representations of Lake Kariba. Once the largest reservoir in the world, Lake Kariba spawned a literary and artistic soul-searching among the colonial population, as it grappled with the contradictions of artificial nature. A white population of 100,000 produced more than thirty books—as well as countless films and works of art—about this single landscape feature. Arriving in Trinidad, then, I expected images and texts on oil everywhere. Surely, a nation of 1.3 million would represent its landscape of rigs, seascape of offshore platforms, and ubiquitous burning of oil and gas in cars and factories. Initially I found nothing. Art and music—which abound in Port of Spain—often depicted nature, more often showed the human body, and focused in particular on the annual Carnival celebration. I found mere mentions of oil and gas in a handful of calypsos. Scrunter's ballad "Oil in the Coil" (1985) associates petroleum with virility and, indeed, with an aphrodisiac quality of men from the petroleum region.[2] More chastely, Earl Lovelace, Trinidad's national writer, penned one line in a play: "With gladness beating in your heart, like them Texaco machines pumping oil out of the earth chest" (1984, 3). I followed up this metaphor of petroleum and vitality, but the trail ended there. I met many musicians, writers, and artists who all agreed on this petro-silence. Some mentioned Trinidad's national instrument: in the 1930s, oil workers fashioned barrels into the steel pan. Again, though, the beneficiaries of this upcycling focused on the container more than on the contents (Campbell 2014, 53). Oil itself fertilized a garden of symbols where almost nothing grew.

This strange sterility has more to do with oil than with Trinidad. Across the world, a century and a half of petroleum production and consumption have imprinted the arts and literature relatively little. In absolute terms, of course, there are many films and texts about oil. Analysts of the humanities mostly prefer to see this glass as half full. Hannah Appel, Arthur Mason, and Michael Watts refer to a "rich loam" for literature. However, they privilege moments "where the normal and calculated course of energy events is interrupted" (Appel, Mason, and Watts 2015a, 10, 14). Introducing another important collection, Ross Barrett and Daniel Worden forgo their own nuanced understanding of "oil's signature cultural ubiquity and absence." They turn quickly to "spectacle" as a central theory (Barrett and Worden

2014, xvii, xxiv). Other observers—with whom I agree more—find hydro-carbons to be blatantly missing in action. It is "startling," writes critic Rob Nixon, "that not since [Upton] Sinclair's California saga *Oil!* [1926] ... has any author hazarded writing the great American oil novel" (Nixon 2011, 73). Nixon cites a "dramatic deficit": oil appears less frequently in culture than one would expect given its economic importance. The Indian novelist Amitav Ghosh diagnoses a dearth of "petro-fiction" and "the muteness of the Oil Encounter," as he terms the social shifts accompanying petroleum (Ghosh 1992, 30). Likewise, Gustavo Luis Carrera begins *La Novela del Petróleo en Venezuela* somewhat deflatingly with, "This book relates to a novel that does not exist. And in that there is no exaggeration. One does not find in Venezuela a fiction of petroleum as, for example there is, in the Hispano-American context, a fiction of the Mexican revolution."[3] A petrostate, Carrera argues, scares writers into self-censorship. Ghosh might agree, but he diagnoses another lacuna in the social relations of oil pro-duction. The oil town—in the Persian Gulf or elsewhere—draws workers from myriad countries. The resulting amalgam congeals too little to form a community that might be narrated. As a final explanation for the scarcity of oil novels, Peter Hitchcock advances omnipresence itself. "Oil's saturation of the infrastructure of modernity," he argues, "[obstructs] its cultural rep-resentation" (Hitchcock 2010, 81). Oil flows like the unremarked air that industry and consumer classes breathe every moment (Huber 2013, 26). Here is a theory of absence rather than ubiquity: state power, social chaos, and sheer familiarity all suppress oil fiction.

To these three explanations I would add a fourth, more technical con-sideration. Petroleum inhabits geological rather than human or medical spaces. Some bitumen, the heaviest hydrocarbon, has seeped into public sight at Los Angeles's La Brea tar pits (LeMenager 2012). Much more oil circulates through middle-class life encased in plastics and vehicles. But the raw, undisguised substance almost invariably passes unseen from sub-terranean strata to enclosed pipes and tanks. One can easily confuse the contents and the container. The photographer Edward Burtynsky, for in-stance, titles his 2009 collection *Oil*, although the images show very little oil (Burtynsky 2009; Szeman and Whiteman 2012). Except for views of the tar sands in Alberta, the photos frame derivatives: pumps, pipes, re-fineries, roads, cars, tires, planes, and ships. Crude itself does not appear. A consumer injects gasoline blindly, without even glimpsing the liquid.

I.1 Sebastião Salgado, "Greater Burhan Oil Field, Kuwait," 1991. © Sebastião Salgado. From Contact Press Images.

Only the abnormal event—the spill—brings a black goo into view and into contact with human flesh, usually the worker's flesh. The most famous photographs of oil itself—taken by Sebastião Salgado (1993, 338–43) in his *Workers* collection—show men plugging wells and fighting fires set by Saddam Hussein's government upon leaving Kuwait (figure I.1). Oil coats their clothes and their bodies.[4] Still, it doesn't become part of them; petroleum washes off.

Coal, on the other hand, operates surgically on the human body. The greatest novel of coal—Emile Zola's ([1885] 1968) *Germinal*—refers continually to the physiology of the French miner. The old man Bonnemort "spit black," explaining, "It's coal. . . . I have enough of it in the carcass to warm myself until the end of my days."[5] He and his coworkers refer proudly to the cuts on their backs—made by low roofs in tunnels—as "grafts."[6] Finally, as a sabotaged mine collapses upon the workers, Zola describes it as "an evil animal . . . that had swallowed so much human flesh!"[7] People enter the earth and the earth reciprocates by giving them silicosis. Diesel fumes can also trigger childhood asthma, but many other contaminants cause that pathology. Black lung is coal's signature. That hydrocarbon, in other words, conducts a "social life," made possible by the "intercalibration of the biographies of persons and things" (Appadurai 1986, 22). Oil lives alone in a studio apartment.

This contrast between the world's two major fossil fuels runs right down the middle of Upton Sinclair's oeuvre. The famous American anti-industrial muckraker penned *King Coal: A Novel* in 1917 and *Oil!* in 1926. Both stories proceed in the manner of a bildungsroman: the young, naive, male protagonist gains knowledge and maturity, specifically discovering and then attempting to ameliorate the lot of the working class. A trio of characters surrounds this hero: his father, a captain of the given industry; a lovely, flighty girlfriend belonging to the same upper class; and a decidedly poorer female with a heart of gold. The hero jilts the princess for a life of activism with the proletarian woman. So closely aligned in cast and plot, the novels differ mostly in their descriptions of the commodity and the labor it entails. Sinclair's petroleum novel introduces readers to the oil field by narrating a gusher: "The inside of the earth seemed to burst out through that hole: a roaring and rushing, as Niagara [Falls], and a black column shot up into the air . . . and came thundering down to earth as a mass of thick, black, slimy slippery fluid . . . so that men had to run for their lives" (Sinclair 1926, 25). In *King Coal*, the equivalent passage—positioned almost exactly at the same point in the novel—describes a more prosaic, but deeper engagement with geology: "The vein varied from four to five feet in thickness; a cruelty of nature which made it necessary that the men . . . should learn to shorten their stature. . . . They walked with head and shoulders bent over and arms hanging down, so that, seeing them coming out of the shaft in the gloaming, one thought of a file of baboons" (Sinclair 1917, 22). Oil provokes flight while coal calls the very species into question. Later in the same passage on mining, Sinclair refers to the colliers as "a separate race of creatures, subterranean gnomes" (1917, 22). Men adapted to the shafts and tunnels. Writing slightly earlier—and in the wake of Charles Darwin—H. G. Wells imagined colliers evolving into a separate population. In *The Time Machine* (Wells 1895), Morlocks—a pun on "mullocks," a contemporary term for miners (Stover 1996, 9)—hunt down the insipid descendants of the rich. In other words, this habitat—which one historian denotes the "mine workscape"—exerts powerful, mostly negative effects on *Homo sapiens* (Andrews 2008, 123–25). Where coal acts continually and viscerally, oil only bursts forth in rare frenzies.

There is one exception, however. In Nigeria, oil has provoked a moral response in literature and more widely as well. Into the delta of the Niger River, petroleum has spewed and spilled prolifically for the last half century. Nine to thirteen million barrels enter marshes and mangrove swamps every year—an annual spill equivalent to the 1987 Exxon Valdez disaster (Baird 2010). There, hydrocarbons break into view, as the sheen on water and as flames flicking from a ruptured pipeline. A photographer like Ed Kashi can capture women baking tapioca by the heat of horrifically toxic gas flares (figure 1.2; Kashi and Watts 2008, 20–23). The dystopia deepens: delta residents attack oil installations, sabotage pipelines, steal oil, and resell it in an extensive network of traders, insurgents, and extortionists (Gelber 2015; Timsar 2015). Oil, in short, busts out of its containers, triggering what geographer Michael Watts (2001) terms "petro-violence," intense struggles over the myth and reality of unearned wealth. Nigerian writers—mostly unknown outside their country—have fashioned these conditions into a genre of "petro-magic realism," laced with themes of indigenous animism, "monstrous-but-mundane violence," and oil pollution (Wenzel 2006, 456). Wealth erupts in spectacle (Apter 2005). At the same time, a palpable "oil doom" prevails in representations of that region (LeMenager 2014, 135). In short, this oil does not behave in anything approaching the conventional fashion. In Nigeria, the economy and infrastructure of oil malfunctions and even collapses. Meanwhile, crude generates all the morally rich meanings so absent in other oil regions. Nigeria is the exception—the anomalous element—that proves the rule of oil's overwhelmingly banal, amoral interpretation.

Elsewhere, hydrocarbons slip into popular discourse almost as unremarked as a cliché. The phrase "black gold," for instance, exerts little critical leverage anymore, if it ever did. That metaphor for money runs through the brief canon of fiction and critical nonfiction on oil in the second half of the twentieth century.[8] Iran's petroleum, writes the journalist Ryszard Kapuściński, "squirts obligingly into the air and falls back to earth as a rustling shower of money" (1986, 347). In Edna Ferber's *Giant*—the only U.S. novel to rival *Oil!*—Texas crude simultaneously enriches and debases the cowhand Jett Rink. He is "touched by the magic wand of the good fairy, Oil" (Ferber 1952, 412). With similar irony, Abdelrahman Munif's *Cities of Salt* (1994) focuses on the overwhelming aesthetic of unearned wealth. The American oil company throws a party on the beach that stuns the

1.2 Ed Kashi, woman baking tapioca by gas flare, Nigeria, 2008. Courtesy of Ed Kashi via VII Photo Agency.

locals: "Sorrow, desires, fears, and phantoms reigned that night. Every man's head was a hurricane of images, for each knew that a new era had begun" (Munif 1994, 221). Finally, in Venezuela, petroleum symbolizes "uncontrollable powers . . . seen primarily as a form of money" (Coronil 1997, 353). Beyond the orbit of these well-known literary and academic texts, financial meanings operate as dead metaphors. Dead metaphors—which might be thought of as merely sleeping—do connect ideas but not in a way that provokes outrage (Kövecses 2002, ix). Oil stimulates the stunted emotion Stephanie LeMenager calls "petromelancholia." Authors of this genre express "the feeling of losing cheap energy" (LeMenager 2014, 102). What about the feeling of, by contrast, using lots of energy of the most ecologically expensive sort? Recall the unprecedented clarity and power of Al Gore's film, *An Inconvenient Truth*, released in 2006. "The moral imperative to make big changes is inescapable," he intones at the beginning. Then, having elevated himself to the top of the hockey-puck curve of CO_2 concentrations, he concludes, "If we allow that to happen, it is *deeply* unethical" (Gore 2006, emphasis in original). Gore then spoke of obligation and a need for restraint. His film reached millions of Americans, but it was not enough to attach conscience lastingly to oil.

Conscience centers on alternatives—on options rejected in the past, options available to us now, and the overlap between these categories. Regarding energy—defined broadly as the capacity to do work—Trinidad presents such a field of actual and possible plans and fantasies. The earliest and most potent alternatives do not involve oil at all. In 1498, during his third voyage, Christopher Columbus sailed through the Gulf of Paria and the 11 kilometer strait between the island of Trinidad and what is now Venezuela. From Orinoco River sediment—visibly discoloring the gulf—he inferred a continental land mass. And land meant an energy platform. To his mind, terrain in the tropics functioned as a kind of solar collector. Rays hit the ground vertically—and not always beneficially. Renaissance geography classified latitudes south of the Tropic of Cancer as a "torrid zone," dangerously hot and sun scorched. That heat created potential too: Leonardo da Vinci classed the sun as a "generating power" (quoted in Mollat 1965, 93). Columbus seems to have agreed with the Italian. After his fourth and final voyage, he averred, "Gold is generated in sterile lands and wherever the sun is strong."[9] Intermittently over the next two centuries, Europeans returned to the region looking for the city of that gold, El Dorado (Naipaul 1969). Not until the 1730s and 1740s did a Spaniard—or one who left a considerable enough written record—detect a different potential in the Orinoco sun. The Jesuit Joseph Gumilla proposed developing a solar colony: a tropical settlement that would thrive on Spanish-planted cacao pulled upward by abundant rays from the nearest star (Gumilla [1745] 1945, 43–47; Ramos Pérez 1958). Today, we refer to this light, heat, and photosynthesis as merely "passive solar energy," incapable of doing work in the mechanical sense. Eighteenth-century theory treated energy more broadly, as a life force, that could inhere in matter both organic and nonorganic (Illich [1983] 2009, 13). Trinidad's sunlight, then, constituted an energy system both local and divine.

And almost immediately forgotten: a half-century later another Spaniard imagined energy and the capacity to do work in very different terms. Josef Chacón took up Trinidad's governorship in 1784 and was the first to succeed in that position—until the English conquest of 1797. Like Gumilla, he sought colonists to grow an export crop, sugar in this case. Mathematically minded, Chacón calculated the inputs necessary for agri-

cultural productivity. His figures omitted sunlight entirely while enumer-
ating slaves in great detail. How many bondsmen were needed per unit of
land, Chacón constantly asked, while seeking to import this labor from
elsewhere in the Caribbean. He recruited settlers—largely French planters
disaffected with the governance of their islands—as a means to acquiring
their human property. What he could not obtain regionally, he tried in vain
to import directly from the African coast. Chacón did not employ the term
energy. Yet plantation slavery and the Middle Passage propagated a new
understanding of that category: no longer as a diffuse life force and not
even as human labor but now as an expendable, consumable fuel. "Arms,"
as the men and women were called, crossed the ocean in the hold of ships.
Buyers and sellers measured them in units, stored, used, and—as they died
from overwork in Trinidad—replaced them. Their agriculture depended
on the sun, of course, but planters devoted little attention to it. In this shift
of values, energy lost both its anchor to certain tropical landscapes and
its divine quality. Chacón, having never read Gumilla, did not appreciate
his own turn from the sacred. He did, however, wrestle with the practical
and moral difficulties of objectifying women and men. At times—as when
slaves fled from their plantations—he had to acknowledge the free will
and all-too-human qualities of "arms." Chacón, then, did not quite achieve
what he, gropingly, set out to do: to establish a pipeline of interchangeable, impersonal energy units. Chapter 1 considers Chacón's successes and
his ethical challenges, scruples that, of course, culminated eventually in
Emancipation.

After Chacón and after Emancipation, another European converted
hydrocarbons into an energy form truly without conscience. Trinidad
contains the most prolific seep of petroleum in the world. Heavy asphalt
literally bubbles to the surface. Indigenous people and Spaniards had
used the black goo for caulking ships and similar tasks. Could one burn
this substance? By the early 1860s, Conrad Stollmeyer—a German immigrant to Trinidad—had distilled the material into kerosene and was
selling it as an illuminant. In 1866, Walter Darwent drilled the world's first
productive oil well in the south of the island. But Stollmeyer—unlike any
other figure in this drama—knew indirectly of Gumilla and his ideas of
solar energy. Indeed, the German had proposed and planned a utopian
colony to be powered by sun, wind, and other tropical forces. God-given
powers, he hoped, would replace not only plantation arms but all forms of

hard, manual labor. This utopia failed immediately and abjectly. Then the German discovered combustible petroleum. In this interval, Stollmeyer juggled all the major energy options—solar, wind, somatic, and petrolic—in his eager hands. He had an ethical choice to make, but—by that point disillusioned with utopianism—he appreciated only its business aspects. Through actions more than words, he married oil with human labor in a fashion that emancipated no one. As chapter 2 narrates, Stollmeyer's loss of conscience helped craft an energy without conscience. Retrospective observers refer to this sort of conjuncture as an "energy transition," a slow but definitive flip from one source to another (Smil 2010, vii–viii). Reading history forward and in its context, however, one cannot pinpoint a flip in Trinidad. Stollmeyer and his contemporaries hesitated as they sorted through immeasurable opportunities and risks.

I want to reconsider that moment of doubt from an ethical perspective. The Caribbean had already witnessed reprehensible acts of breathtaking proportions (Khan 2001). Europeans had virtually wiped out the islands' indigenous people, only to replace them with enslaved Africans and indentured Asians. Capitalism, racism, and Christianity all contributed to extraordinary violence. But—alongside and partly independent of these forces—a new idea of fuel took hold. In Trinidad, producers and consumers of energy came to see it as a transportable, interchangeable commodity. This ideological and moral shift has never figured among the famous transformations of the Caribbean—or of anywhere really. Trinidad's historiography tends to treat oil and gas merely as substances and as unalloyed goods for the island and beyond (Mulchansingh 1971; Ministry of Energy and Energy Industries 2009). In both world wars, Trinidad's oil propelled British and Allied forces. After Independence in 1962, the country developed its gas sector, becoming a major exporter of downstream products such as methanol and plastics. Oil has given the country economic stability and political sovereignty. Thus, thanks to relatively open governance and technical competence, Trinidad has largely skirted oil's frequent "resource curse." The specters of underdevelopment, corruption, violence, and pollution do haunt the island. But the Orinoco delta is no Niger delta of oil theft and paramilitary politics. Trinidad's hydrocarbons appear to have solved many problems without creating substantial new ones. *Energy without Conscience* seeks to overturn that comforting account. Trinidad—like any state producing or consuming hydrocarbons—must reckon with the contemporary great evil of dumping carbon dioxide in the skies. True, the effects of burning oil have taken longer to accrue than did the earlier body counts of Atlantic conquest or capture. But damage now becomes more evident each year. The historical part of this book (part I) returns to the 1780s and 1850s, when solar, human, and fossil energy sources seemed simultaneously promising and problematic. Revisiting the paths not taken, we might discern a better choice.

Complicity

I have struggled to find a language with which to describe the varied conditions of my informants in Trinidad. Like many of us, they burn hydrocarbons at rates higher than the global per capita average. The women and men of this first group of Trinis drive cars, live in air-conditioned houses, and use energy in all the ways characteristic of the world's billion high emitters. Many of my informants go further than that: they control private firms and government agencies that exploit hydrocarbons systematically. This second group comprises "captains of industry"—in the quaint phrase used without irony in Trinidad's convention halls and luxury hotels. A third set of informants captains nothing, not even motor vehicles. The residents of South Trinidad's oil belt consume little oil. They become relevant to this story because of their choice not to protest the oil and gas industry. The practices I describe then range from promoting oil, to reaping its benefits, to remaining silent about its costs. Environmentalists might describe the first party as responsible for climate change and the last one as ignorant of it. Perhaps the consumers in the middle—for whom we still lack an adequate descriptor—act negligently toward the atmosphere and everything dependent on it. If climate change were solely an environmental problem, then this lexicon would do the job: I would present the ethnography of people variously enabling one form of pollution. But I don't consider climate change to be merely an environmental problem. It is that and much, much more. The commodity chain from hydrocarbons to hurricanes—which I treat as one unit—has occupied the land like a far-reaching system of power. Combustion, as Rob Nixon (2011) writes, wreaks a "slow violence" as devastating as it is pervasive. Occasionally, a fast Pakistani flood or Louisiana hurricane causes death tolls too high to measure with accuracy. Some authors describe this uneven lethality as "petro-dictatorship" or "fossil capitalism." Climate change thus exceeds other ecological crises in both its scale and its delivery of force. I am less concerned with labeling this system than with understanding those operating within it. They are, I argue, "complicit" with oil.

In this sense of widespread but traceable, anthropogenic harm, colonialism may provide the best analogue.[10] Almost as total as climate change, the system of rule prevailing over the Americas, Africa, Oceania, and much of Asia for as many as five centuries contained fast and slow violence. Around 1800, outright enslavement and genocide gave way to Christian and other "civilizing missions." European scientists began an "anticonquest" of discovery and description. The geographer and explorer Alexander von Humboldt contributed more than anyone to this movement. His and contemporary texts, though, could not avoid complicity. So writes Mary Louise Pratt, charging various narrators with constructing "cultureless" brown and black bodies available for European domination (Pratt 1992, 53). Pratt may have indicted von Humboldt unfairly (Marcone 2013), but she indicates the difficulty any intellectual faces in thinking outside the dominant ideology of his or her time. In the twentieth century, though, the colonial paradigm began to crack. In 1937, George Orwell denounced both imperial working conditions and left-wing intellectuals' tolerance of the same: "In order that England may live in comparative comfort, a hundred million Indians must live on the verge of starvation—an evil state of affairs, but you acquiesce in it every time you step into a taxi or eat a plate of strawberries and cream" (1937, 159). This charge—holding a large but defined group responsible for vast harm—could just as well apply to users of fossil fuels today. One can no longer plead ignorance. The information that, say, carbon emissions are pushing millions of Indians into starvation and displacement is widely available and credible. To choose the car over the bicycle, one has to repudiate science. Few people reject climatology explicitly. Far more high emitters deliberately discount or refuse altogether to imagine current and future victims of climate change. That decision takes place almost, perhaps entirely, automatically, but it constitutes a discrete action: "acquiescence," in Orwell's turn of phrase. Small, prosaic actions are beginning to accrete to the level of mass death.

At that larger scale, with whom does the accomplice conspire? *Complicity*, which shares a root with *accomplice*, implies a partner in crime.

Perhaps oil serves as the trigger man. Bruno Latour (2005) might put the argument in these terms: networks of human and petrolic "actants" collaborate on the basis of complementary properties. The harried commuter, in other words, wants to reach her destination, the motor vehicle carries her, and the petrol pushes the piston. More recent scholarship focuses on the vibrant quality of materials, as if gasoline willed people from suburb to suburb and jet fuel flew them personally from continent to continent. Certainly, energy behaves in ways that suggest volition (Bennett 2010, 54). It moves at the speed of electrons or explodes into atoll-destroying mushroom clouds. Many of my informants in Trinidad credited oil and gas with an understated animacy. Deposits were constantly welling up, and, as chapter 3 explains, petroleum experts portrayed themselves as hardly more than helpmates to the nearest gusher. Such modesty actually shifts responsibility to the hydrocarbons themselves, as if humans only lately joined a geological plot hatched elsewhere. Ethnographically, I treat such theories as a folk belief—or folk science—that obscures political and economic relations. On the ground, *people* populate the network that wills carbon emissions—and, therefore, climate change—to happen. Producers collaborate with consumers to move oil from underground reservoir to refinery, to engine, to atmosphere. Almost all the time, that process unfolds exactly as the sentient actors intended, anticipated, or could have anticipated it to do. Hydrocarbons are an instrument, like the hammer that one uses to pound a nail into a piece of wood. Until something goes wrong: oil does—let's say—conspire against people when its volatility causes a refinery to explode and contaminate the local environment. The CO_2 spill everywhere, on the other hand, figures only as the last link in a well-functioning commodity chain designed and operated entirely by men and women. At opposite ends of a long pipe, consumers act as the party complicit to producers of oil, and vice versa.

That multiplex human partnership encompasses only some people, some societies, and some states. The bulk of our species—minus the one billion high emitters—participates in oil mostly as victims of it. I do not share the mounting concern that humanity has become a geological agent, ushering us into the so-called Anthropocene era. The chemist Paul Crutzen popularized that neologism in 2002 to indicate "mankind's growing influence on the environment."[11] By now, a wide range of scholars, journalists, and activists defines the Anthropocene as "the first geological era shaped by one species, humans." That charge assumes an onset of the Anthropocene from the domestication of plants or from the Pleistocene extinctions caused by the first Native Americans, as if maize cobs led inevitably to megatrucks (Ruddiman 2013). A minority of *Homo sapiens*—"industrial humans" perhaps—developed hydrocarbons and everything they power. Today a minority dumps gigatons of carbon into the atmosphere (Malm and Hornborg 2014). True, almost everyone buys plastic and other products containing oil and transported by burning oil. Yet the Zimbabwean peasant who lights her mud-and-pole dwelling with one petroleum-based candle hardly counts. She practices what Anna Tsing (2012, 95) calls "slow disturbance," artisanal lifeways that mostly recraft biodiversity. The prefix *anthro* spreads blame too widely (Chakrabarty 2009, 216). A small guild, so to speak, manufactures lethal climates for mass distribution.

In focusing on that guild, I have written a customary sort of ethnography. Part II of *Energy without Conscience* examines the current life of tribe-sized, faraway social groups so as to illuminate problems in North America and Europe. The bulk of my readers, I suspect, live—as I do—in the Global North and consume hydrocarbons at a fast clip. My informants live in Trinidad and Tobago and engage with hydrocarbons in additional ways. But the cultural distance is not so great that I need to familiarize you long-windedly with my subjects. The particular hurdle for this book lies in describing some of my informants as unusual at all. Crude oil, as the term even suggests, is ordinary, pedestrian. To disrupt that normalcy, the activist Bill McKibben labels oil, gas, and coal firms as "radicals. They are willing to alter the chemical composition of the atmosphere in order to get money. That's as radical an act as any person who ever lived has undertaken" (Climate One 2011). Trinidad and Tobago's energy experts find petroleum and gas where no one else does, and some of them export their knowledge to Africa and elsewhere.

Despite this trail of damage, I do not consider such people monsters, motivated by hate or beyond the arc of reason. My informants practice their professions in a fashion that both benefits society in the short term and uses a natural resource that would otherwise be neglected. They contribute only complicitly to a project larger than themselves. To that project, additional clusters of Trinidadians contribute less directly. Chapter 4 concerns environmentally minded activists, some of them poor and undoubtedly low emitters. These men and women became complicit by omission: they refused to protest the global oil spill, as well as local ones, and in so doing crafted a narrow, indeed obsolete, politics of pollution. Finally, chapter 5 discusses what I call the climate intelligentsia of Trinidad, a loose group of scientists, activists, and policy makers who portrayed Trinidad as an innocent victim of climate change. Astonishingly, their rhetoric of small, vulnerable islands exonerated the country's oil and gas sector. These individuals all held erroneous assumptions, a fact that most—and mostly with humor—acknowledged to me. Some are now trying to move Trinidad's own energy grid from gas to renewables. Most, though, want simply to produce another barrel of oil.

The Feeling of Energy

How does it feel to change the climate in sensory, rather than moral, terms? Feeling connotes tactile experiences as well as ethical dilemmas. The former do not immediately lead to the latter. To take things in proper order then—as an ethnographic subject lives her life—let me ask, "How does it feel, in sensations, to consume energy?" Matthew Huber has already probed this issue in relation to U.S. suburbs. They present "an appearance of atomized command over the spaces of mobility, home, and even the body itself" (Huber 2013, 23). People feel free, as they flit in cars between detached houses and points of consumption. Residents of Port of Spain, or at least of its wealthier parts, also know this behavior and its sense of liberation. Many wake in the middle- and upper-class fringes of the city and travel into or through the urban core daily by car. I followed this pattern, sometimes alone and more often sharing transportation. The daily journey covered what one might call three energy zones related to different objects: automobiles, bodies, and buildings. Port of Spain is what Carola Hein (2009) calls an "oil capital." But it also seems to pulse with something more elemental—a kind of mania and revelry in the consumption of energy per se. Cars, exercising men and women, and air-conditioned edifices huffed and puffed visibly, even promiscuously.

The first sensation comes with combustion, the thrum of engines, and the pull of g-forces. With my family, I lived in Cascade, on the fringe of Port of Spain. We rented the house of Eden Shand, a retired politician described at length in chapter 5. As the name suggests, Cascade slides down the foothills of Northern Range, off dramatic ridges and into steep ravines.

1.3 Port of Spain viewed from St. Ann's. Photograph by the author.

The vistas are beautiful—and mostly accessed by car (figure 1.3). In recent years, developers have built roads and houses at the very limits of the automobile. Vehicles will not ascend slopes steeper than those in Cascade and its adjoining settlement, St. Ann's. The landscape then turns commuting into something more intrepid and exciting. I rode sometimes with Che Lovelace, as he descended Cascade with my son, with other children, and with eight long boards for a Saturday surf lesson. We whizzed through sinuous, riparian curves, the sea peeping through dense foliage, as Che drank a shandy or talked on his cell phone. Elsewhere we might get stuck in a traffic jam. But in Cascade driving was fun, and people reveled in it. Cheap gasoline—subsidized by the petrostate—underwrote this automobility. But a feeling enlivened it. Perhaps it was the thrill of driving in an urban geography not quite meant for the car, as shown in the foreground of figure 1.3. To me the lanes always seemed too narrow, the curves too blind, and the gradients way too up and down. In this sense, Cascade differed from a safe, sedately mobile American suburb. The car in Cascade—as it burned petroleum—pulled one up, down, and sideways.

At its southern apron, Cascade and St. Ann's spill into what I would call a zone of body energy. The Queen's Park Savannah, the greensward in the middle of figure 1.3, separates downtown from the northern outskirts. On that very grass in 2007, Eden Shand deployed his body against the car, protesting the paving of a southern section of the Savannah. A truck dumped gravel on him, damaging his spine permanently. Around the Savannah runs a 4-kilometer sidewalk, which is Trinidad's closest approximation to a pedestrian mall. People don't merely idle and stroll. Fit women and men come to see and be seen as they expend energy. Most go clockwise, with the car traffic, and no one crosses the Savannah. Running shoes on, I sometimes took part in this crowded rush hour of muscle and movement. It peaks in January, as people methodically tone their bodies for Carnival. They are enacting a cosmology—with a more positive outcome than in Shand's case. In Trinidad, writes anthropologist Daniel Miller, "the truth of a person exists in this labour they perform to create themselves" (2011, 50). Those exertions bear fruit as near-naked bodies cross through the south stands—along the same Savannah edge—to be judged on Carnival Tuesday. I "played mas," as they say, dressed as a bare-chested pirate. With my wife and two friends, we "chipped" down the road from sunup to sundown for two days. I believe there is no outdoor recreational event where so many people work so hard under such equatorial heat for so long. Rio's Carnival takes place mostly at night. The Boston Marathon finishes in a few hours. In Port of Spain, masqueraders sweat like slaves, practicing an art form derived from slavery. But even as they expend somatic power, they do not feel anything like slaves. At the edge of the Savannah, where a parking sign instructs, "four taxis facing north," I ran into the author of a short story by that name (Walcott-Hackshaw 2007). She was dancing with herself, with her body, blissed out and oblivious to the world.

That taxi rank marks the boundary of Port of Spain's third energy zone. Elites have built an archipelago of air-conditioning. From the point where I saw the writer in rapture, one crosses Queens Park West Road into the neighborhood of Newtown. Once a frontier of urban expansion, these dense blocks contain headquarters of foreign-owned oil companies: BP, Repsol, EOG, and British Gas. I did not go into these edifices very often. My research centered on Trinidadian firms and organizations. But I wandered those streets, sometimes meeting informants in the Rituals Café on Marli Street. Even outside one feels the energy of cooling. Frigid air pours out, unimpeded by double doors or any of the other energy-saving methods employed elsewhere. Businesspeople emerge from buildings overdressed, scurrying from the tropical heat into climate-controlled cars. The Guyanese novelist Oonya Kempadoo (2001, 17) describes a look of "air condition skin," conveying wealth and the habit of self-protection from the elements. Perhaps a whole neighborhood can wear this aesthetic. Trinis themselves remark more frequently on the air-conditioning of another locale, about a mile south of Newtown. On the Gulf of Paria, the government had recently established an International Waterfront Centre. Its Hyatt Hotel and two glass spires — in the right background of figure 1.3 — deliberately evoked Dubai. The Ministry of Energy and Energy Industries occupied some office space, but most of the square footage stood empty. Trinis joked about governmental hubris and speculated on air-conditioning. Dry season temperatures exceeded 90°F every day for months. Was the state burning its natural gas reserves to cool vacant acres? Or was it letting them bake, and risking equally expensive damage to the buildings? Workers at the ministry understood more than the average person about heat and energy. One usually burns fuel to *raise* temperatures. There is something miraculous — always seemingly futuristic — about combustion for cooling. It involves more artifice and people know it. Certainly, energy executives — with their "air condition skin" — knew it as they hurried from one vessel of privilege to another.

I conducted most of my ethnography along this energy-intensive transect of motors, muscles, and manipulated air. In Cascade, I lived near some of my informants, but not with the close immediacy of the classic peasant or tribal study. "Studying up" — as we call the ethnography of elites — requires surmounting barriers against access (Nader 1974). Petroleum geologists live behind walls, in gated communities. I had to meet them over lunch, over drinks, or in their offices. Conferences allowed me to carry out true participant observation. There — often in the resplendent Hyatt Hotel — I joined discussions and receptions with the most accomplished and powerful energy experts. To be objects of anthropological study alternately flattered and amused them. As I pushed this indulgence, attitude became my method. Promoters of oil and gas are wrecking the world. This conviction — my feeling about energy — has driven this study from the beginning. Initially coy, I gradually deployed this sentiment. If you really care about sea level rise, I would say over rum, why don't you just leave the hydrocarbons in the ground? It was a provocation reminiscent of the filmmaker Michael Moore (2004)—who, in one memorable scene, asks congressmen to enlist their children for military service in Iraq.[12] Moore did not amuse his interlocutors. Perhaps because Trinidad has a tradition of teasing—called *picong*—energy experts took my jibes in stride. They laughed and then responded revealingly. Still, I wanted more. I wanted to find someone who agreed with me. So I left my customary corridor in Port of Spain and explored the oil fields and industrial sites of South Trinidad. I found people opposed to pollution in their communities, and asked, "Would you really be satisfied if this industry left here merely to export harm elsewhere, possibly to the whole planet?" Most would have been. Again, I learned a great deal while gaining little peace. I found data but not truth as moral clarity.

At least, I found complex individuals: the planner-cum-slaver Josef Chacón, the utopian-turned-oilman Conrad Stollmeyer, the eco-driller Krishna Persad, the selective environmentalist Wayne Kublalsingh, and the lady-doth-protest-too-much prime minister Patrick Manning. Throughout the book, I attend to the consciousness of key figures in the energy trade. Many of these men—men have consistently dominated the energy sector—failed in their own terms. They imagined more than they invented. Conditions frustrated their ambitions, or they themselves sold out their loftiest ideals. Why should any living or long-dead leader with few followers then attract followers now? The question or criticism would seem all the more pertinent in Trinidad, which has developed a tradition of cynical, distrusting appraisal. Eric Williams, the country's historian turned independence leader and first president, rose to prominence by debunking the pious sentiments of British abolitionists. Bondage was unprofitable before it became unpopular, he wrote in *Capitalism and Slavery* (Williams 1944), rather than the reverse. In the same spirit, Trinidad's talented calypsonians revel in unearthing corruption. Ridicule eventually touches every politician. Trinis understand complicity all too well. Meanwhile, anthropology has never privileged the individual over the collective, or the singular insight over the idea widely shared. In writing something like biographies, then, I am cutting against a grain of local discourse as well as the disciplinary sensibilities of my own social science. It is necessary to do so. Or, rather, my political agenda—to challenge people's complicity with climate change—compels the most thorough search for precedents and examples of life without fossil fuels. In this sense, I resist the label "utopian." Everything that I and others seek in energy has already happened to someone or to someplace.

In *Energy without Conscience*, I am trying simultaneously to amplify and dampen public concern about climate change and fossil fuels. As a cause for alarm, this commodity chain threatens the conditions for life on planet Earth. Bill McKibben (2010) designates the carbon-enriched environment "Eaarth," a new spelling for a new, profoundly dangerous age. By 2100—if business continues as usual—grain belts will collapse and ecosystems will have already hemorrhaged species. Ecologically speaking, nothing this bad has occurred since the last mass extinction event, the dinosaur-destroying meteor strike of 65 million years ago. This time, as many observers now quip, "We are the asteroid" (e.g., McKibben 2003, 11). Here I would qualify the alarm and its misguided universalism. The "we" of seven billion *Homo sapiens* has not acted in concert. As a set of deeds, climate change is spreading in a patchy, discontinuous fashion. Environmentalists see this pattern every day. It is a planetary version of the toxic risks and exposures concentrated in poor communities of the Global North and South. Burning oil constitutes a form of environmental injustice and human-on-human structural violence. This interpretation—suggestive of war—indicates helpful, sober precedents. The United States devotes a fifth of its government budget to defense. More than a million men and women relinquish their liberty to serve as soldiers, sailors, and airmen. Imagine shifting all those resources and goodwill to defending ourselves from oil and climate change. If motivated by a national emergency, high emitters would replace oil and gas with wind and solar, conserve energy, and live differently. Perhaps considering oil as a merely military threat will help us phase it out (Garrard 2004, 107).

Thus, I would like to decelerate and redirect the rhetoric of apocalypse. Apocalypse, by definition, arrives without precedent and requires unprecedented defenses and adaptations. To relinquish fossil fuels, for instance, might require a dictatorship or "climate Leviathan" capable of repressing consumer choice in high-emitting democracies (Wainwright and Mann 2013). That speculation exceeds the bounds of this history and ethnography. *Energy without Conscience* contributes to the debate, nonetheless, by suggesting that people have already envisioned the abandonment of oil. I do not share Slavoj Žižek's (2010, 334) despair in writing that publics imagine the end of nature more easily than the end of fossil capitalism. For that latter event, many societies have already trained and know—if only through their historical archives—more or less what to do. Trinidad once planned development without oil. There, in the eighteenth century, a Jesuit designed an agriculture powered only by equatorial sunlight. The governor of Trinidad harnessed the power of African bodies. Both schemes imagined what we now call alternative energy. A historian—or one narrowly tethered to chronology—might consign these failed plans to an ash heap of impractical or immoral attempts. As an anthropologist, I have (or have taken) the liberty of running history backward, excavating the solutions that predate problems, and indulging in counterfactual speculation: what if people had not banished God from the landscape, or what if, from the wreckage of Caribbean slavery, survivors had salvaged the value of walking, pedaling legs as useful energy? From off the favored Euro-American stage, this study engages in what Svetlana Boym (2008, 4) calls "off modern" thinking—"an exploration of the side alleys and lateral potentialities" of where we are.

There may be no better way to approach the question posed at the outset of this introduction: How does it feel to change the climate? How, furthermore, does it feel not to care? Where, I might add, is conscience, or guilt? Where—and this is what I also mean by *conscience*—is a sense of responsibility or reverence for energy and the world around it? McKibben wrote in 1989 about living morally with "the end of nature." He awakes into an "alertness," akin to the tensing of a swimmer hearing a distant motorboat (McKibben 1989, 49). McKibben's unease mounted so high that he founded the first climate change movement in the United States. I would like us all to acquire the same fear and to respond with a measure of McKibben's desperation and generosity. My informants stand at quite some distance from this position. From petroleum geologists to antitoxic activists, they mostly don't care deeply about climate change. They care now and then, but they don't care about global warming in that way that one worries over a sick, elderly relative, growing feeble, losing capacity, heading for a different state. Perhaps no one cares about climate change in the way that that senescent person herself faces mortality and the uncertainty of what lies beyond. The absence of those feelings presents a shape. It has contours and boundaries. The ethnographer, in conversation with someone vaguely concerned about climate change, brushes against the skin of that silence, provoking defensiveness, a glance of recognition, or a joke that both parties know is not funny (cf. Kidron 2009). As much as nonfiction can do, *Energy without Conscience* attempts to illuminate that negative space. Let us see not-feeling-climate-change as a concrete thing. It sits among us like an antiquated superstition, too customary to discard but too backward to celebrate. I wish to expose that belief as retrograde and wrong. With this historical and ethnographic story, I hope to crack the chalice of disregard still cradling oil, its producers, and its consumers.

PART I

ENERGY *WITH* CONSCIENCE

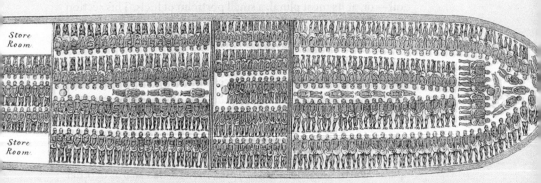

Trans-Atlantic slaving ship

FROM ENERGY TO FUEL, so much has been lost. Anthropologists frequently lament the attrition of languages, religions, and cultural customs of all sorts. These aspects of culture live on, but frequently as curiosities and performances for tourists. How many people dance only for cameras in hotel lobbies, selling revelry, lust, or anger as a global commodity? These practices still mean something—possibly more than before—but they lack conscience. Commerce has hollowed a thickness and density once palpable around us. Energy has thinned in the same way. What was a multiplex notion of divinity, life, and rightness now denotes oil—or, at its most plural, a small portfolio of fuels. This section of the book traces that narrowing: from solar power to slavery (chapter 1) through a last consideration of both of these possibilities before the final choice of hydrocarbons (chapter 2). Energy now fills a tank. In the United States, traces of the former holism still remain: the energy in a yoga studio or the energy of New Age crystals. But these references bulk small: they are hollow in the sense that they hardly offset the overwhelming drivers of what has been called fossil capitalism. Modernity runs consistently on a planet-destroying fuel. I am hopeful, though, that shards of other cosmologies—recognized and again respected—will inform energy policy.

CHAPTER 1

Plantation Slaves, the First Fuel

By 1700, the sugar revolution had taken hold just north of Trinidad, on Barbados. Another British colony, Jamaica, was also enjoying a boom, as did the French isles of Martinique, Guadeloupe, and Dominica. Most infamously, Saint-Domingue, later Haiti, worked Africans to death by the thousands. Plantations soon festooned the Caribbean, pumping out sweetness and wealth. With the notable exception of Trinidad: the island languished as a colonial failure. Spain neither knew what resources existed on the island nor garnered investment capable of exploiting more than a fraction of them. A handful of settlers and a small population of slaves planted cacao only to suffer a devastating crop blight in 1727. Five years later, the Jesuit Joseph Gumilla visited for two weeks and recommended rehabilitating the crop: Madrid, he envisioned, would send landless peasants from Andalusia and the Canary Islands to grow cacao without slaves.[1] Solar rays and the natural fecundity of the tropics would guarantee good harvests. But the Crown sent no one. Then smallpox struck in 1739 (Joseph 1838, 148). The colony teetered on the edge of ruin. "Even the monkeys died," writes V. S. Naipaul, whose novelistic sensibilities capture the era better than straight history. "The morale of the settlers broke. For a century and more they had lived close to nature. Now, ignoring the Spanish code, they left their huts in Saint Joseph and lived, like the Indians before them, in the bush" (Naipaul 1969, 123–24). Civilization teetered on the brink of collapse, a defeat all the more bitter given its Caribbean context. For the region's easy money, Trinidad possessed sunshine and fertile soil but lacked every other necessary element: wealthy settlers of the planter class, equipment for sugar mills, and, above all, slave labor. Although no one phrased it in quite that way, Trinidad suffered from a crisis of somatic, or bodily, energy. For a late sugar revolution, the island needed manpower, measurable in kilowatt-hours but sought then by the boatload.

Into this breach strode a man shrewd and calculating enough—and just barely ruthless enough—to succeed. In 1783, the Crown and its *intendente* (administrator) in Caracas appointed Don Josef Maria Chacón, a naval brigadier and knight of the Order of Calatrava, to serve as governor of Trinidad. He was to implement the royal edict, or *cédula*, of 1783: a set of reforms intended to encourage planters with slaves to immigrate to Trinidad. Contemporaries described him as polyglot, indefatigable, and incorruptible (Joseph 1838, 160–61, 168). In these qualities, he probably did not differ substantially from Gumilla. But, in order to implement the cédula, he disregarded—or, at any rate, did not dredge up—the Jesuit's writings. Chacón took as given the need for labor and, in particular, for unfree labor. This somatic project, however, created its own problems. The governor struggled with the task of objectifying workers. Enslaved Africans expressed themselves and acted as individuals, especially when they absconded from plantations. Even as he sought shipments of human cargo, Chacón could not easily repress the personalities contained below decks. Still, he did so well enough to turn the corner in Trinidad. The colony began to produce, refine, and export sugar—at nowhere near its potential—but at a level respectable enough to attract attention. London noticed. Thirteen years into Chacón's governorship, England captured the island from Spain without firing a shot. Chacón, who might otherwise have retired in glory, instead returned to Madrid in disgrace.

To the extent that he did recruit planters and slaves to Trinidad, Chacón provoked the most profound energy transition of all. He and other traders of slaves invented fuel. A fuel stores energy in a measurable, countable, transportable, and salable form. Energy becomes fuel as it becomes a resource. But resources—such as lakes or forests—need not move. I write *fuel*, then, to emphasize this intrinsically deracinated quality. Solar rays bore none of these attributes. Gumilla never thought to package sunlight and send it from sunlit to shady areas or from the tropics to the temperate zone. Laborers, even slaves, did not automatically assume this commodity form either. Some served the master and his family over a lifetime, acting as acknowledged persons in a social field. Other slaves—particularly in the context of plantations—performed the tasks assigned day in and day out with no personal recognition from above. Marx wrote of these people as possessing "labor-power" but without the proletarian's right or ability to sell it. Whereas workers might advance by learning new skills, plantation

slaves would gain little or nothing. Lacking incentives to specialize, they remained or became general, interchangeable, and substitutable (Marx 1976,
1032). Plantation hands became liquid, one might say. Traders transported
labor power over far greater distances than they had done for wood, the
Caribbean's closest approximation to a modern fuel. Cane cutters, in short,
helped Chacón and his successors to imagine energy for the first time as a
commodity and as a flow. But by running away and even killing their masters, they also constantly challenged that understanding. At moments like
these, Chacón had to consider them as individuals: a prick of conscience.
The first fuel, thus, flowed imperfectly, slowed by the friction of moral scru-
ple. It flowed well enough, nonetheless, to establish the conventions under
which we now extract oil and ship it across oceans by the boatload. Without intending to do so, Chacón and other sugar revolutionaries imagined
the true energy without conscience that waited in the wings.

Scientific Slavery

Trinidad entered the slave trade as a scrounger. Spain lacked the West African ports and interior networks so useful to Portugal, Britain, and France.
The empire could only obtain Africans through intermediaries and after
those parties had satisfied their own needs. Trinidad's status in the empire—as a backwater of a backwater—limited the options further. How
could remote Port of Spain, Trinidad's capital from 1757, direct the flow
of enslaved Africans to its shores? Scavenging seemed like the only option. In 1763, as a result of Europe's Seven Years' War, Britain acquired the
French Antilles of Dominica, Saint Vincent, Grenada, and Tobago. Catholic planters stayed in place but chafed under Protestant, foreign rule. Incentives might induce them to leave—and bring their slaves and expertise
to Trinidad. In that expectation, the Crown relaxed its highly protectionist
controls on imports in 1776. Some Frenchmen island-hopped to Trinidad,
among them a certain Grenadian planter, Philippe Rose Roume de St. Laurent. Roume immediately sought to recruit more like him. He estimated a
further 379 families were available on Grenada and on Martinique, where
ants were ravaging the cane crop. They would bring 33,322 slaves in tow
and require, Roume estimated, one 3-*fanega* plot of food crops per group
of seventy.[2] (A fanega comprised roughly 7.5 acres, or 3 hectares [Joseph
1838, 162].) These stipulations would, as he put it later, guarantee "the es-

tablishment of the Colony of Trinidad and means to bring it promptly to perfection."[3] Prompted by Roume, the Crown proclaimed the cédula of November 24, 1783. More generous than the exemptions of 1776, these "Regulations of Commerce and Population" excused Catholic settlers from import duties, from the annual taxes, and from other financial burdens (Newson 1976, 179–80). Slave masters would receive a basic land grant and another 50 percent of that acreage for each bonded man or woman imported. With an odd precision, the incentive scheme awarded $2\frac{1}{7}$ fanegas per slave.[4] The subsidies worked, and planters—whether driven by desperation or greed—began to arrive (figure 1.1). Trinidad might finally become a sugar isle.

Chacón intended to make sure of that, indeed, to blanket the island with plantations. After a delay, he arrived in Port of Spain in September 1784. Possibly the governor already possessed a familiarity with quantitative measures concurrently applied to estates in Saint-Domingue. There, the plantation owner and geographer Moreau de Saint-Méry was in the process of establishing a ratio of two hundred slaves for 150 tons of sugar (Martin 1948, 122). More broadly, French Physiocrats were enthusiastically—and for the first time—systematically counting everything from births to trees (Scott 1998, 14). In this intellectual movement, Chacón played a small, local role: he perfected a quantitative, scientific approach to slavery. He began by facilitating trade—or, as the instruction from the imperial intendente in Caracas put it (with no irony intended), "to liberate the slaves forever from all import duties, in light of the increase that will result in agriculture."[5] How much increase would result? Chacón spent his first month calculating the relationship between somatic energy, botanical production, and financial rents. He treated the land as if it were a giant machine. Reporting in November 1784 to Caracas, he worked backward from the surface area of Trinidad, estimated at 400 square leagues, of which 180, or roughly 100,000 fanegas, were arable. Chacón borrowed yield figures from French and British isles, while anticipating that Trinidad's "tierras virgenes" would produce more per fanega than did the more settled Antilles. Based on prevailing agricultural prices, he then compiled the ideal crop mix for the plantations, and, most importantly, the number of slaves required per fanega. The governor derived employment rates for sugar as well as coffee, cotton, and cacao. Projections of profit favored cane and hard work. But the labor intensity of cacao and coffee–$1\frac{1}{2}$ and $2\frac{1}{2}$ slaves

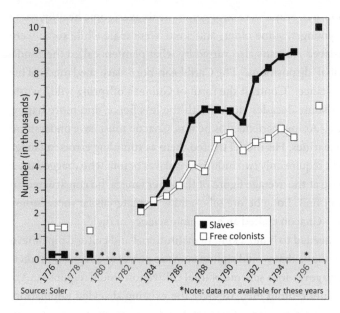

1.1 Settlement of Trinidad, 1784–1797. Created by Mike Siegel of Rutgers Cartography Lab.

per fanega, respectively—indicates Chacón's deeper insight: one could measure slaves in fractions—as increments of labor power—rather than as whole bodies.[6]

This calculus completely overlooked the messy realities of plantation life. Planters knew better. Sugar barons on Saint-Domingue could not begin to calculate a stocking ratio of labor: they rotated the workforce among fields and crops, presumably exploiting the change of seasons and maturation of individual cultivars (Bonnet 2008, 147–48). Even within a field, difference frustrated calculation. In Barbados—the archetypical Windward sugar island—plantations frequently intercropped sugar with food provisions, such as maize and sorghum (Roberts 2006, 579). Presumably, such polyculture reduced the efficiency of cane harvesting but enhanced the overall productivity of the land. Thus, the sugar revolution involved more species and more complexity than that catchphrase suggests. In other ways as well, Chacón and all counters of slaves were chasing a myth. Barbadian planters aimed for a labor density quite close to Chacón's ideal—one slave per two acres—but they knew that slaves differed (Curtin 1990, 83;

Menard 2006, 93). Some were stronger, some weaker; some healthy, some sick; some younger, some older; and some were male while some were female. Enslaved Africans also varied by what planters called their individual or tribal "dispositions." The Caribbean-born slave shed much of his parents' resistance. "Contrast the form of Guinea's offspring wild," wrote a Trini poet some decades later, "With the less fierce, and more happy, Creole child."[7] As inexperienced as he was, Chacón must have understood these particularities. His model treated them as rounding errors: a fanega of cane might require only two industrious slaves or four who dragged their feet. To arrive at the average figure of three per fanega, he constructed an abstract labor unit. Long before oil—and with materials far more lumpy and unruly—Chacón crafted an economic science of fuel.[8]

Then he needed to build a kind of institutional and cultural pipeline: a set of conventions that would siphon somatic power at a regular, predictable rate. The governor's scheme of total cultivation would require 234,373 Africans, a hundredfold increase over the existing shackled population. Roume revised his estimate of importable slaves upward to 40,000—all "acclimated" and accompanying masters immigrating from other islands.[9] Even this highly optimistic projection left a deficit of nearly 200,000 labor units. Port of Spain would need to procure bodies directly, in bulk, and possibly unacclimated. In 1784, therefore, the governor initiated a business relationship with Edward Barry, the Havana-based agent of a Liverpool firm (Newson 1976, 217). Chacón struck a good bargain. On October 4, the Englishman disembarked 584 slaves, expecting payment only with the upcoming January harvest. The labor units, Chacón anticipated, would amortize their cost in one season. In a matter of days, however, their productivity fell to zero. "The climate surprised them," Chacón reported on October 8. Lacking shelter and even hammocks, they slept on the humid ground, where many sickened and died. Admitting his own failure to provide food and lodging, Chacón reloaded the survivors onto Barry's ship and forwarded them to more capable buyers on the mainland.[10] Then, in November, Chacón contracted for slaves resident in Cuba, workers who would presumably tolerate the climate better. But this trader, referred to as Fitch, ran into difficulties with the first shipment. His frigate sailed with only 437 of the expected 500 slaves—and after an expensive delay. Trinidad still lacked the facilities to accommodate a larger influx. Chacón wrote his superiors in February 1785, "We would not have been able to comply with the agreement without great losses, had [the House of Fitch] delivered, over the last four months the 4000 slaves it offered."[11] The island could hardly import and maintain all the somatic power it needed to apply.

Still, in the same letter, Chacón doubled down on an increasingly risky pipeline bet. "Whatever European trading house succeeds in selling 4 to 6000 Negros annually at the price of 750 pesos each," he argued, "can just as well guarantee a number of 7000 slaves paid at the end of the first year."[12] Trinidad, he concluded, would have to graduate from Caribbean scavenging to large-scale, transatlantic hunting. "It is one of the tasks that occupy me incessantly," he reflected in 1787. "The slaves from the part of the coast of Africa frequented by the Portuguese," he continued, acknowledging differences in disposition, "are preferable to those brought by the English, French and Americans, as much for greater docility as for more skills and strengths."[13] Chacón sought loans with which to purchase human cargo (Soler 1988, 34). In Caracas, however, the intendente preferred to assist with tax breaks, having lowered import duties again to "attract colonists who have slaves and funds to build houses and dedicate themselves immediately to the cultivation of the land."[14] In fact, the colonists dedicated themselves immediately to clearing land, creating a further problem for the economy of slavery. Lowland and montane tropical forest covered all but a fraction of the 100,000 fanegas slated for plantation agriculture. As Frenchmen disembarked, they set their slaves to work cutting primary growth, probably the hardest labor in the Caribbean (James [1938] 1989, 56). Already in 1784, Chacón reported a death rate of one in three forest clearers.[15] The inflow of slaves was not sufficient, Chacón worried, to "replace the number of those who have died."[16] In 1788 alone, 893 lost their lives (Brereton 1981, 27; Noel 1972, 99). Chacón blamed tropical forest and, more specifically, "sicknesses caused by the first exhalations of lands that had perhaps not seen the sun since they left the hands of the Creator."[17] A more humanitarian—or simply more conservationist—governor would have encouraged masters to work their chattels less. But Chacón already thought like an oil company. He treated bonded people as a substance to be obtained and consumed. Rather than restrain consumption—as fuel became scarce—he hunted for new supplies.

Somatic power might walk on four legs rather than two. In 1790, the governor began petitioning Caracas for mules. Hooves were scarcer than arms: five years earlier, Chacón had reported only fifty-seven of these animals working in agriculture (Noel 1972, 105). Mules, of course, could not hold a machete to cut cane. They could, however, turn the mill that crushed stalks and squeezed out their juice. As the sugar crop expanded in the 1780s, this processing step began to limit production. Caracas proved more useful in freeing up this bottleneck, and mules numbered more than a thousand by 1794 (Soler 1988, 117). Still, in 1796, Chacón underscored "the losses that [colonists] suffered this year for lack of mules to mill the canes and carry out the other tasks of their estates."[18] Compounding the problem, mules dropped dead from too much work and too little feed at an astonishing rate of 15 percent per year (John 1988, 16; Newson 1976, 200–201). Chacón himself was suffering from a hoof-focused tunnel vision. Of 135 mills, only three operated by water and one by wind. The Trades blew constantly over the Lesser Antilles, turning sugar mills on Antigua, among other islands (Sheridan 1972, 22). Why did Chacón not seek to replicate that energy system? Why did he not, as was the practice in other Caribbean islands, position water wheels on the many streams flowing off the Northern Range and into the environs of Port of Spain (Patterson 1967, 55)? Chacón's correspondence reveals no inkling of these place-based possibilities. Perhaps mules succeeded well enough to validate his fixation on muscle. As far as human, somatic fuel went, 6,431 slaves worked 1,463 fanegas at the peak (Soler 1988, 114). Chacón boosted sugar cultivation but came nowhere near realizing his megamachine. Shortfalls, however, did not undermine his most lasting contribution: the embryonic idea of an amoral fuel, coursing throughout the Caribbean. Whether empty or full, a pipeline now connected Trinidad.

Slaves as People

Trinidad needed a constant influx of slaves not only to enlarge the supply but just to stay even. Two forms of offtake drove the population slowly down. Slaves died—worked to death—and slaves achieved their freedom. The former circumstance both resulted from and contributed to the treatment of people as labor units. The most oppressive estates handled Africans as consumables. Fortunately, this style of management did not last. After the British interdiction of the slave trade in 1807, plantations took much greater care to ensure that bondsmen reproduced and sustained a local population. They treated at least female slaves more like breeding cattle than like the sterile, and therefore expendable, mules. Even in Chacón's time, conditions for slaves improved. In 1789, Spain adopted the Code Noir of revolutionary France. The new rules required masters to baptize and evangelize among their labor force, to encourage marriage, to facilitate the cohabitation of spouses, and to minimize physical abuse (Joseph 1838, 174–76; Newson 1976, 200). Trinidad's planters were not pleased. Still, they manumitted nine hundred slaves between 1784 and 1797, an astonishingly high rate for the Caribbean (de Verteuil 1992, 174). Perhaps the limits to exploitation also eroded the benefit of owning some slaves. Or slaves gained recognition as people. "I have been moved to promise them freedom," wrote Don Herbert Roy of his mistress and her child in 1793, "because of the great love I have for them" (quoted in de Verteuil 1992, 178). Such emotion cut to the core contradiction of slavery: the difficulty of managing, as an insensate object, a woman or man who obviously felt, thought, and acted (Davis 1966, 261). Chacón operated in this murky space—on the edge of conscience.

Some slaves exerted their will in ways the governor could not ignore. For want of soldiers, his predecessor had failed to control roughly three hundred maroons, or runaways, hiding in the forest around Port of Spain (Newson 1976, 200). More than 10 percent of the chattel population had, in other words, taken their labor elsewhere permanently. In common with the authorities on other islands, Chacón paid prizes to the captors of slaves. The resulting manhunt itself stimulated laborers to turn themselves in. Chacón delivered these individuals back to their original masters, advising the latter to watch them closely. "[I have] destroyed entirely the pack of fugitive Negros," he boasted to Caracas in late 1784, although the island was not yet "entirely calm with its slaves restrained and secure."[19] Indeed, by 1788, the problem had reappeared with a vengeance: Chacón feared incipient "cantons" of absconded property. "I propose to his Majesty," he wrote, "the establishment of a company of Dragons [cavalry and infantry combined] . . . sufficient to attack and collect the fugitive Negros."[20] He was asking a lot: How could an island then barely able to import mules expect to obtain warhorses? To underscore the need, Chacón raised the specter of "the English and the Dutch [who have felt] the need to recognize fugitive slaves as independent, buying through shameful treaties a peace they could not achieve through force and in which, by consequence, they cannot secure trust."[21] Still a military man at heart, the governor feared maroon ministates of the sort carving out territory in Jamaica and Surinam (Bilby

2005; Thompson 2006, 296–97). Conscience would not restrain him in preventing this outcome.

What was really at stake in the question of maroons? Each arm exiting the sugar fields lowered production and depleted the assets of the grower. The maroon also modeled an alternative way of life. The original Spanish term *cimarrón* referred to cattle as well as people. As they crossed the estate boundary, slaves went feral, subsisting on wild foods, on what they could grow while on the run, and probably on food stolen from the plantations themselves. With the exception of this last method—which Chacón did not mention in a letter until 1788—maroons made the landscape productive under conditions of freedom.[22] Much as Gumilla had proposed, they lived off sunlight and tropical fecundity. Chacón and his associates did not make this comparison, but they understood how much maroons hurt the colony's credibility. The governor's initial instructions from Caracas referred to *negros cimarrones* as causing "grave damage to owners, and to themselves, since, as soon as slavery is shaken off, they wander errantly and are prone to laziness and vice, from which follow pernicious consequences."[23] At the time, the damage appeared to be mostly symbolic. Maroons demonstrated that slaves did not need masters. Worse than seceding as a ministate, self-liberating men and women implied the possibility of a full-on revolution, of exactly the sort that ignited Saint-Domingue in 1791. Ironically, Roume found himself there, serving as a commissioner of the French government (itself undergoing a revolution). A mob captured him in 1795, keeping him for nine days in a chicken coop (Korngold 1944, 189). *Marronage* could lead to this: master and livestock traded places, and Haiti's sugar plantations evaporated into thin air.

In short, the project of imagining slaves as energetic objects almost—but not entirely—succeeded. Circumstances forced Chacón and slaveholders repeatedly to consider slaves as people—and, therefore, to apply his conscience in dealing with them. Roume got out of the chicken coop by signing a decree demanded by the crowd. Whether through violence or entreaties, enslaved humans demonstrated their capacity to act. They contributed song and dance to Trinidad's earliest Carnival celebrations (Liverpool 2001, 127ff.). More fundamentally, enslaved Africans and their enslaved children made personal histories and characters. Such individuality would disrupt any commodity. One can only trade, say, bushels of grade A wheat if every bushel carries exactly the same qualities as every other. If every slave worked at the same rate every day, then the master could reliably stock his fields with three per fanega. Laborers would function like barrels of sugar or, better yet, as wood used to heat cane juice to a boil: they would serve as the faceless fuel of the plantation machine. No commodity plays this role without the constant investment of institutions and power. Through calculation and planning, Chacón tried to insist upon a standard, but he came to understand how slaves varied. Bonded bodies crossed oceans and seas as "lively commodities," an uneasy mix of standardization and wildness (Collard 2014, 153). In the latter capacity—sometimes as lovers and sometimes as loafers—they defied the measures of volume and ratios of production applied to them. As a fuel, slaves carried somatic power. As an imperfect fuel, they also exerted the force of their personalities.

Did the concept of fuel demand this shearing away of meaning? "Certain forms of knowledge and control," writes the political scientist James Scott, "require a narrowing of vision" (1998, 11). The trader of oil cannot sell it efficiently as long as he or she must record and advertise its place of origin and vintage. A vintage adds value to wine but can do the opposite for packages of energy, especially in a high-energy society. One cannot design, say, a fleet of Cadillacs to run on the particular sour crude of Trinidad and Tobago. Simply put, fuel functions as a commodity in the contemporary sense of the term: a crop or mineral traded globally by a standard unit of measure. Perhaps, then, Chacón accomplished something akin to what Karl Polanyi (1944) calls "the Great Transformation," the disembedding of land, labor, and numerous goods and services from their social context. I would not go this far. Anthropologists now dispute Polanyi's claim, finding embeddedness in everything, even in money (Maurer 2006). Energy lost something more specific: its landscape. The Middle Passage displaced solar power and deracinated somatic power. Partly as a consequence, the ability to do work also lost a certain enchanted quality associated with the sacred. For Gumilla, sunlight carried virtue and sprang from God, "El Autor de la Naturaleza." Neither slavers nor consumers of oil have tended to treat these forms of energy as God-given. One *could* do so, however. Perhaps, I would speculate, we treat oil as mundane for the same reasons Chacón preferred to treat enslaved African as arms. Standard measures narrow vision enough to obscure sin. The slave trade destroyed individuals and African societies. Even before climate change, the oil trade poisoned individuals and polluted societies, many in Africa. Consumers have always been able to investigate this pain. To do so, though, requires detective work at the point of origin. The trace—the vintage of this harm—does not travel with the commodity across the seas. Fuel's great transformation disembedded energy from ethics.

Energy without conscience, then, accompanied the rise of capitalism. The relationship is not straightforward. Plantation slavery arose before and outside the industrializing core of Europe. Industrialism ultimately rendered transoceanic slavery and the plantation form obsolete. Sugar estates innovated too little, perpetually leaning on the crutch of cheap energy (Carrington 2003). Yet Chacón and his contemporaries did not see themselves as stagnant. Catching up to the seventeenth- and eighteenth-century sugar revolution, the governor and his French planters aspired to run "factories in the field" (Davis 2006, 104). The Caribbeanist Sidney Mintz (1985) famously described sugar plantations as a satyr-like amalgam of agrarian and industrial forms. In a fashion we now associate with Henry Ford's assembly line, they intercalibrated men and materials as a metabolic system. Herein lay the lasting innovation of plantation slavery: a cultural understanding of production through long-distance, high-volume energy transport. This idea need not have eventuated in capitalism, but global capitalism would not have flourished in the twentieth century without it. Now, of course, solar panels and wind turbines can generate electricity for industry and other uses nearly anywhere. But in Chacón's time—long before the electric wire—production needed its first pipeline. It was an idea as well as an infrastructure. Chacón hardly articulated the idea. His calculations simply demonstrated its necessity, requiring no further justification. In this way too, fuel avoided a moral reckoning. Shallow questions of infrastructure and logistics predominated and still do. Doggedly, unreflectively, Chacón filled boats with bodies. Barrels would not be far behind. Trinidad was ready for oil.[24]

CHAPTER 2

How Oil Missed Its Utopian Moment

Oil could have developed differently. Absent some contingencies, the substance might have entered history as a moral category—at least, in Trinidad. As a burgeoning set of hopes and ideals, energy with conscience lay within grasp. In the middle of the nineteenth century, the term *energy* was gaining currency in Europe, defined as various forms of the ability to do work (Coopersmith 2010, 264ff.). With dreams, experts might fashion energy into a tool for reform, liberation, or justice. Or would the first combustible hydrocarbon serve only as a fillip for production? At the very birthplace of the oil industry, Trinidad and early Trinidadians wavered between inspiration and indifference.

The man who tipped the scales was the island's most influential German immigrant. Born in Ulm in 1813, Conrad Friedrich Stollmeyer emigrated to the United States, where he became an outspoken abolitionist. In the 1830s, he read the French utopian socialist Charles Fourier, who considered all hard labor to be slavery. To harness the human body, Stollmeyer concluded, was deeply immoral. Embracing leisure, he pledged himself to an ideal then gaining currency as the "paradise without labor." Tropical fecundity and solar power would do the job of muscle and sinew. To prove the practicality of this substitution, Stollmeyer moved to Port of Spain in 1844. From that base, he organized a utopian scheme across the Gulf of Paria in Guinimita, in newly independent Venezuela. Stollmeyer attracted working-class emigrants from Britain. But they fared far worse than had Chacón's Francophones: fourteen of the thirty-seven died within months, and the colony collapsed utterly. Despite this debacle, Stollmeyer remained in Port of Spain and continued to ponder work and energy. In the 1850s, he gained a position as manager of an asphalt deposit in South Trinidad. With his partners, he found a method to distill that heavy hydrocarbon into a light oil that would burn and generate heat. He had, at last, found a reliable substitute for human bodies and a means to the paradise without labor. Yet chance and historical contingency intervened. In Port of Spain, Stollmeyer observed freed slaves not laboring—enjoying their leisure—and he was appalled. His sentiments flipped entirely: work did not enslave men, he now felt, but improved and invigorated them. His oil alleviated no toil in the plantations. Instead of sending it there, he sold it for illumination, fueling the first streetlights of Port of Spain. As an emancipator, Stollmeyer failed because he no longer wished to succeed.

Petroleum failed too. Oil might have obviated the need for mass labor in agriculture and manufacturing. Of course, Trinidad, along with the rest of the British Empire, abolished slavery in 1838. France freed its slaves in 1848, and the laggards—the United States, Cuba, and Brazil—followed in the last four decades of the nineteenth century. Freedom, however, did not abolish drudgery. Men, women, and children continued to labor in the hot Caribbean sugar fields and in European factories. In theory, the dense bonds of hydrocarbon molecules could have replaced people at harvest time and before the loom. Scholars describe such a substitution as an "energy transition" or a change of "energy regime" from somatic to fossil sources (McNeill 2000). But the change did not unfold that way—not in the Caribbean and probably not anywhere. Rather than displacing one another, energy sources have tended to accrue. A new fuel usually allows industry and agriculture simply to produce more goods (York 2012). Already in 1922, the American critic George Santayana wrote, "inventions and organization which ought to have increased leisure, by producing the necessaries with little labour, have only . . . degraded labour and diffused luxury" ([1922] 1968, 192). In short, oil could have paid a popular dividend of ease, but oilmen—including Stollmeyer—diverted those benefits. Oil's reputation might have profited from slavery's disgrace.

Imagine a counterfactual in which refined petroleum directly substituted for unfree labor. Oil and the people who drill it would emerge as great emancipators. Given the timing, events might have played out in this straightforward fashion. Stollmeyer distilled asphalt, the heaviest hydrocarbon, into flammable kerosene in the 1850s. He could have burned his way into history as a great emancipator. But the German wanted none of this glory for himself or for his product. He didn't actually like black people. Abolition, in his view, enabled resegregation: the return of Africans to their own shores. Perhaps because of this racism, he crafted a postslavery, post-somatic energetic unit without celebration. Oil flowed as an impersonal, placeless liquid. It aroused no moral outrage—and no moral endorsement either. In this sense, Stollmeyer persevered exactly where Chacón and other slavers eventually faltered. He founded a fuel that would power enterprise while passing through port unprotested, mostly unremarked altogether. In the longer term, he bequeathed to oil a symbolic flatness—an absence of conscience—that would obstruct any political reckoning. Although an interesting figure then and now, Stollmeyer helped make oil boring.

Slaves, Iron Slaves, and Sunlight

The contingencies marking oil as merely lucrative began with an energy crisis—experienced as a labor shortage. Somatic power still existed, but the British *prise de conscience* had established strict limits on its use. Certainly, one could no longer work African women and men to death. Even the more humane growers of sugar in the Antilles confronted the emancipation of 1838 as a triple calamity. First, as planters now had to buy labor power, wages skyrocketed. Second, labor became less available with each freedman who abandoned the plantations. Many ex-slaves valued independent living—as urban dwellers or smallholder farmers—more than any available form of employment. Third, the terms of trade for British-grown sugar began to collapse. Even after France abolished slavery in 1848, West Indian growers were competing against slaveholders in Cuba, Brazil, and the United States. Inevitably cheaper to produce, slave-cut sugar squeezed the sales of Jamaican, Trinidadian, and Guyanese exports. In possibly the first fair trade movement, Anglo-Caribbean growers pleaded with Parliament for the application of an import tariff or boycott against slave-made sugar—but to no avail. London seemed to want to have its cake and eat it: that is, to enjoy the moral satisfaction of freeing slaves without paying the resulting higher prices for a sweet cup of tea. This set of transatlantic debates became known as "the sugar question." Under that title, a widely circulated 1845 pamphlet proposed what seemed like the only solution implementable by West Indians: the shipment of free immigrants across the Atlantic to the sugar islands. "Let Africans populate our West Indian Colonies," exhorted the writer, "until Guiana is as dense as Cuba" ("The Sugar Question" 1845, 17). Such a migration promised both to civilize African laborers and ensure the continuation and financial soundness of emancipation. In the event, Jamaica, Trinidad, and Guiana (now Guyana) did import cane cutters—from India rather than Africa and as indentured, not free labor. In retrospect, Britain answered the sugar question by reinventing slavery in a milder form.

In 1845, Conrad Stollmeyer—who dreaded this outcome—proposed a third way between slavery and immigration: the "iron slave." Unique among commentators on the sugar question, Stollmeyer defined the problem as a shortage of energy, rather than one merely of human beings. Workers, free or chained, did not possess sufficient quantities of that ability to do work. "It is a law of nature," Stollmeyer wrote, "that man should not submit to more compulsory (by force or wages) labour than he can help, notwithstanding all the preaching against idleness by interested or short-sighted parties" (1845, 17). Born to a patrician, fairly idle family himself, Stollmeyer had become a theist while studying at Stotsingen University. In 1836, he emigrated to Philadelphia, finding work as a publisher of German-language books. There he developed his antipathy toward bondage and joined the Anti-Slavery Society in Pennsylvania (Stollmeyer 1845, 95). By the early 1840s, when he left the United States for Britain, he had formed precise ideas about labor as a whole. He answered the sugar question with his own manifesto, "The Sugar Question Made Easy." The solution, he began, depends "entirely upon the vivifying power of the sun" (Stollmeyer 1845, 7). Metal robots would convert solar power into labor. "ONE IRON SLAVE," Stollmeyer wrote breathlessly, "will do the work of *three hundred* human slaves." "Go to the *iron* districts of England," he exhorted the West Indian planters, "have them [the machines] well moulded and cast . . ." (1845, 18, emphasis in original). The design already existed, sketched by another utopian, John Adolphus Etzler, and published by Stollmeyer in Philadelphia 1841 (Etzler 1841). Armed with a variety of tools, this "Satellite" would cut trees, plant seeds, or harvest crops. In other words, iron slaves promised to resolve the energy crisis.

Yet they were not entirely analogous to human slaves. Here Stollmeyer oversold the Satellite. In an effort to persuade planters, he portrayed it simply as a superior bondsman. "Slaves should be most obedient," he wrote sardonically of shackled humans, "have no will of their own . . . and never grumble or break down during the performance of their work" (Stollmeyer 1845, 17). Such flesh-and-blood automatons did not exist. As Chacón knew so well, slaves ran away as frequently as they performed their jobs reliably.

In the fashion of a true automaton, then, the Satellite would outperform its human equivalents. Etzler described his invention as "imperishable, [and] indefatigable." It would "furnish as much inanimate power as desired, for ever" (Etzler 1841, 1, cover page). This durability marked a crucial difference between organic and metallic slaves. In Trinidad and in the more successful sugar islands, planters worked Africans to death as they imported replacements. The land, one might say, ate slaves, like a machine with its own industrial metabolism. Chacón's "arms," in other words, served as the consumable fuel, not as fixed assets. By contrast, the utopians' iron slave would function as a stable machine or an engine, driven by and consuming solar or wind power. For solar power, mirrors would focus sunlight on a water vessel, producing steam to turn gears. For wind power, enormous sails would rotate those wheels directly, pulling the agricultural machine. As shown in Etzler's drawing, the assembly devoted most of its material and bulk to the conversion of breezes—themselves generated by solar energy—into mechanical movement (figure 2.1). "Iron slave" was a misnomer then. The two men enslaved the sun and the wind while combining the hoe and the machete into a rickety machine.

Etzler and Stollmeyer had met in New York in 1840 at a celebration of the birthday of Charles Fourier (Nydahl 1977, xvii). Both admired the French utopian. Etzler, also German born, had emigrated to Pittsburgh and had been spreading Fourier's ideas through books and tracts (Stoll 2008, 42). From Philadephia, Stollmeyer was publishing the works of Fourier's other American disciples. These utopians took issue with Robert Owen and his brand of agrarian socialism (Brostowin 1969, 266–67). Too much depended on long, backbreaking hours spent in the fields. Work itself was the problem. "Compared to the idleness and well-being which he [the worker] enjoys on Sunday," Fourier wrote, "this indirect form of slavery is not any less physically constraining than real slavery" (quoted in Beecher and Bienvenu 1971, 141). It was not any less dispensable either. Christians, Fourier insisted, were mistaken when they assumed work to be a permanent curse for Adam and Eve's trespass. "Scripture did not say that this punishment would not end one day, nor did it claim that man would never be able to return to the happy state he first enjoyed" (quoted in Beecher and Bienvenu 1971, 149). People could restore Eden by sharing work, dividing each job into short and enjoyable assignments to which they were drawn through "passional attraction." In the community Fourier

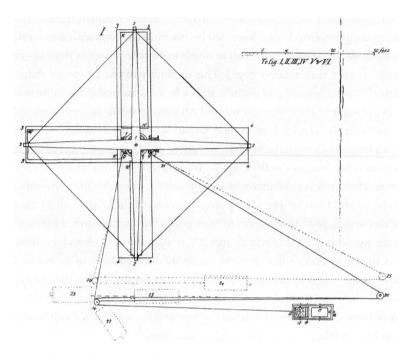

2.1 Etzler's Satellite and windmill viewed from above.

imagined—known as the Phalanx—tasks would range from farming to cooking to light manufacturing, and each member would do many of them for each other in reciprocal rotations (Beecher 1986, 454ff.). This paradise satisfied more desires than did the Christian one. Fourier gave an example of "the young Bastien, [who] to acknowledge Celiante, who has obliged him in various services, will hardly neglect to offer the proof of gratitude that a young man of twenty years can offer to a lady of fifty."[1] From this titillating utopia, Etzler and Stollmeyer deleted carnality, embraced leisure, and identified an energy gap (Claeys 1986). The Phalanx needed a means of capturing inexhaustible power. By definition—given the purpose—they sought energy with conscience.

In that search, the utopians carried out what geographers call a spatial fix. They reached far outside the world of Fourier—to the tropics and their European proponents. In 1804, Alexander von Humboldt had completed his epic journey through northern South America, Mexico, and the Caribbean. He and his companion, Aimé Bonplan, published their account in thirty volumes, the last one appearing in 1834. Considered the founding text of biogeography, this oeuvre mapped the tropics as a series of botanical and climatic zones. Humboldt and Bonplan also owed much to prior European notions of the New World, including those of Gumilla. Like the Jesuit—whom he had read and cited—von Humboldt wondered at the continent's bursting fecundity (Ewalt 2008, 7, 181; Gerbi [1955] 1973, 223n290). His reports of profuse, untamed flora inspired Etzler's 1844 pamphlet "Emigration to the Tropical World for the Melioration of All Classes of People and of All Nations." That document estimated that "twelve to 25 times the present population of the world would find room and food within the tropical zone alone" and "plantains will yield as much nutritive stuff on one acre, as 133 acres of wheat or 44 acres of potatoes in Europe, according to Humboldt, an undisputed authority" (Etzler 1844a, 5; 1844b, 2). Stollmeyer, too, echoed the reference, describing "the luxuriance of the vegetable world in the Tropics which astounded even Alexander von Humboldt" (1845, 15). Sunlight nourished this growth. Under "solar heat," wrote Etzler, "the atmosphere deposits sugar" in wild and domesticated plants (1844a, 6; Brostowin 1969, 237). For mechanical power, he imagined "gigantic powers of nature, of wind, water, and the sea's waves" (Etzler 1844a, 346). At low latitudes, Stollmeyer and Etzler believed they had solved the problem of want.

Etzler relied upon a dizzying method of quantitative extrapolation. As a single-minded planner of agriculture, the utopian outdid Chacón: he blanketed not just Trinidad but half the world with slave-free breadbaskets. The global carrying capacity, asserted in *The Paradise within the Reach of all Men without Labour*, rested on an agricultural yield of food for sixteen people per acre, or 10,000 people per square mile (Etzler 1833, 98). He applied that figure to the entire midsection of the Earth, between 30 degrees North and 30 degrees South latitude, irrespective of deserts and mountains. Rafts made even oceans cultivable, irrespective of hurricanes and typhoons. Sometimes, though, Etzler's enthusiasm snapped his slim tether to numerical reason. Regarding solar power, he began with a notion of focusing mirrors first advanced by Archimedes. A surface of 4 square feet might concentrate solar rays on an area of only 2 square feet. Etzler multiplied this ratio by 100 and then by 100 again to generate "prodigious heat . . . probably greater than any ever known" (1833, 35–36). Etzler then tossed calculation aside in favor of infinity: "We are under no limit for producing any quantity and degrees of heat by this means." Mechanical power would follow. By aiming solar rays at a boiler—and generating steam—Etzler could outdo any existing engine. Again, though, he jumped scale to the cosmos. "Have I asserted too much," he asked, "when promising to show that there are powers in nature a million times greater than the human race is able to effect by their united efforts of nerves and sinews?" (1833, 45). Etzler probably did not want a true answer to his question. Events would soon embarrass him beyond measure. Even Stollmeyer, who only estimated a modest 300:1 ratio in strength of iron to human slaves, failed utterly to demonstrate that capacity in what proved to be an unforgiving corner of the tropics. Still, at that point, his heart and his conscience were in the right place.

The Road to Guinimita

Etzler and Stollmeyer attracted a sizeable and credulous following. Between 1844 and 1846, the Tropical Emigration Society—based in Bradford, England, and led by Thomas Powell—attempted to put the utopians' ideals into practice (Chase 2011). The organization's weekly periodical, the *Morning Star*, served as the rhetorical vehicle. Oddly, though, the magazine's editor, James Elmslie Duncan, devoted very little space to what should have been a chief concern, the feasibility of the iron slave. Stollmeyer and Etzler had traveled to England, and, on September 22, 1845, they conducted a public trial of their prototype. Gadgetry clawed against the ground with little effect and with no energy source besides coal (Stoll 2008, 120–21). In the *Morning Star*, Stollmeyer admitted to "*little* and accidental causes which hindered us from coming off with a grand *éclat*" but affirmed absolutely "the possibility of performing agricultural operations with inanimate powers" of sunlight, waves, and so on.[2] A week later, Duncan assured readers that a Satellite was "made, tried, and efficient for all its purposes."[3] In the new year, Thomas Powell, the titular head of the Tropical Emigration Society, reported, "The Satellite, we are sorry to say, is not yet completed, but will be ready shortly to come to London."[4] There, quite blandly, the matter rested. Meanwhile, on the question of land acquisition, extended, florid articles were describing llanos, mountains, and coastlines, mostly suitable for European settlement. Duncan writes of the "astonishing productiveness of the alimentary plants."

To underscore the thesis of tropical fecundity, the magazine excerpted and adapted Colonel Francis Hall's (1827) monograph, *Colombia: Its Present State in Respect of Climate, Soil, Productions, Population, Government, Commerce, Revenue, Manufactures, Arts, Literature, Manners, Education, and Inducements to Emigration*. Hall had worked as a hydrographer for the government in Bogotá, during the brief postindependence interlude when it administered Venezuela (1823–30). "All the energy of nature in the production of both animal and vegetable life," he exults, "is here [in Colombia] brought into action . . . into a system which man vainly imagines is for *his* peculiar use and convenience" (Hall 1827, 9, emphasis in original).[5] The hydrographer pinpoints parts of northern South America where farmers might avail themselves of nature's energy while conserving their own. Duncan paraphrases Hall's description of Varinas, an inland, piedmont province consisting "entirely of plains . . . [which will] produce abundantly cocoa, indigo, cotton, sugar-cane, tobacco, maize, rice, and all kinds of fruits and vegetables."[6] A handful of other regions offer similar advantages, but *Colombia* includes only one map, of Varinas. Duncan takes pains—even exceeding Hall's text—to warn readers away from the coast: districts "to the northeast extending to the Golfo Triste [the Gulf of Paria] are . . . exposed to inundations, unhealthy, and thinly inhabited."[7] "Black vomit," as yellow fever was known, stalked these shores (Hall 1827, 122).[8] Duncan may have later wished that he had printed Hall's diagnosis in bold in that decisive summer of 1845. Instead, the editor advised, "We must bid adieu to the luxury of linen," in favor of cotton for the tropics.[9] The *Morning Star* treated serious matters trivially and trivial matters seriously.

Perhaps for this reason, two members of the Tropical Emigration Society ignored all the available advice. In November 1845, Thomas Carr and Charles Taylor established a settlement across the bay from Port of Spain. The two men had negotiated for a disused estate named Guinimita (see map 1.1). Writing to the *Morning Star*, they narrated the transaction without describing the acreage itself. That truth only came out as Carr and Taylor assessed the estate's interior. "It is a hilly poor land of no account here," they admitted, "but important for us as a sanatory locality for the members, being dry and breezy, till the flat land of the valley be made more healthy and free from mosquitos by clearing and planting."[10] Their backhanded criticism made the worst part sound the best. When that ominous letter appeared in the *Morning Star*, settlers had already paid their deposits to sail from England to Port of Spain. Powell went with them, and Stollmeyer, who remained based in Port of Spain, coordinated their passage across the gulf. News returned to him as corpses and in the form of letters of explanation. Thomas Carr complained of ailments ranging from perspiration to rheumatism and "weak stomach and biles."[11] W. E. Prescod, another writer and the previous owner of Guinimita, reported himself fit as a fiddle. In twelve years at Guinimita, he and his family had never been sick. "I challenge the Tropics," he declared against all geographical explanations, "to produce a healthier race than what is on it." Settlers had been "indiscrete," he explained. One died of sunstroke. A second overheated and, in an attempt to cool himself at night, lay nude on the cold ground. Fatal fever followed. Most bizarrely, Prescod diagnoses "burden of 'black vomit,' from eating to excess rich American cheese."[12] Did settlers debilitate themselves by overindulging in Guinimita's few pleasures?

Neutral observers found fault in the organizers of Guinimita. Stollmeyer absorbed most of the published attacks. Like Prescod, he accused the victims and exonerated all other factors. "The climate is certainly not to blame for the imprudence of any people," he argued in the *Morning Star*.[13] Stollmeyer told his side of the story in Port of Spain too, and the *Gazette* of that city responded with a caustic exposé, reprinted humiliatingly in the *Morning Star*. Settlers, the newspaper revealed, had been reduced to subsisting on flour and water. Stollmeyer resembled "a man who took it into his head that he could teach his horse to live without food and who had got him down to a straw a day," whereupon the horse died in what the man called an accident.[14] The *Gazette*—considered the planters' newspaper (Cudjoe 2003, 134)—clearly believed that Europeans could acclimate to the tropics. But the process took care. Stollmeyer, they opined, had neither exercised such prudence nor seemed to have learned it subsequently, believing still in "the perfect practicability of European colonization in the tropics." Indeed, his excuses, the paper said, "calumniate the dead as well as disgust and shock the living."[15] Ironically, one reader of the *Morning Star* had predicted the mortality in a fashion both precedented and prescient. In that fateful summer of 1845, a letter suggested "succinct rules for the guidance of the society in the choice of a locality for the settlement of the Colony." Avoid moist ground, the author urged, as "it will be found charged with the worst *exhalations*."[16] The miasma-fearing Chacón had been right: tropical nature might kill, just as well as nurture, those who attempted to farm it.

Chacón's slaves actually expired from a combination of overwork and environmental conditions. So did the newcomers to Guinimita. While waiting for the Satellite slave, they used their somatic energy to excess. Between the lines, supporters and critics of Guinimita depicted morbidity through labor. Whitehead, the first to perish, appears to have been doing the laundry of thirty-seven people. Mr. Tucker, who died after evacuation to Trinidad, "had been overexerting himself working twelve to fourteen hours to construct a boat."[17] That half-finished craft is very possibly the same "hulk of a schooner [to which] they [the settlers] could retire for shelter during their labour."[18] Construction of a proper house advanced, "how slowly, with all one's endeavors."[19] The iron slave certainly would have been useful. Utopians must have remembered Etzler's promises of "leisure and freedom from care and drudgery [which] will beget a desire and taste for the refinements of life" (Etzler 1844a, 18). Those refinements might have included cheese. But all the ridiculous and sublime aspects of that vision depended upon an enormous infusion of mechanical, forest-clearing energy. Amid the debacle, few even recalled iron slaves. Was Powell referring to them when he wrote, "Mr. E. . . . lacks the energy necessary to set his own machines going"?[20] The letter is unclear. In any case, in 1845, settlers saw only one energy source that could make Guinimita bloom: themselves. Like fuel, they had crossed the ocean in a boat, duped rather than chained. They could have functioned as energy with conscience, powering an intentional community. But these arms were too few. Stollmeyer's paradise without labor required more labor, toil, and drudgery than he had ever imagined possible. Or it required oil.

Somatic Power Reconsidered from the Vantage of Oil

In the next twenty years, Stollmeyer—still living in Port of Spain—acquired all the means necessary for a nearly labor-free paradise. Having recovered rapidly from the disgrace of Guinimita, he found an energy source commensurate with his utopian conscience. He gained control of the largest natural hydrocarbon seep in the world. The colonial governor, Lord Harris, introduced Stollmeyer to Thomas Cochrane, Earl of Dundonald, a semiretired naval officer, abolitionist, and one-time hero of the British naval blockade against slavers. Cochrane had acquired a concession to extract bitumen, known as pitch, from a deposit in South Trinidad, the Pitch Lake.

He was experimenting with bitumen as a substitute for coal. Mostly absentee, Cochrane delegated these ventures to Stollmeyer. In the 1850s, the two men refined bitumen into the first form of combustible petroleum—and founded the Trinidad Petroleum Company (Higgins 1996, 14). Their work required more ingenuity than one might expect. At the Pitch Lake, hydrocarbons bubbled up steadily and in enormous volumes, but in an extremely viscous form. Sir Walter Raleigh had used this pitch to repair his ships. In the twentieth century, pitch, also known as asphalt, would cover roads all over the world. How could an industrialist convert this heavy goo into a fuel light enough and volatile enough to burn and release energy? By 1860, Cochrane and Stollmeyer had solved that problem through a process of heating and purification. They produced kerosene, which rapidly replaced whale oil as a fuel for illumination (Williamson and Daum 1959, 56). "Pitch lamps" soon cast their wan, flickering glow over nighttime Port of Spain. With a higher energy density than any other nineteenth-century hydrocarbon, kerosene delivered light and warmth in a small package. Conceivably, this fuel could replace somatic power. Here, at last, was the necessary ingredient for iron slaves and an island with less labor. Yet by that point, neither man viewed economic development in remotely utopian terms.

Stollmeyer—whose ideals had been loftier than Cochrane's—had already sold them out in a rapid, well-documented fashion. Even before the dust of Guinimita had settled, Stollmeyer applied for a government position in agriculture. He sought to train otherwise idle freedmen as small farmers. Did he wish to make them work? Perhaps not: his letter of application referred to Trinidad's "most productive soil, which requires *so little labor* to yield abundant food."[21] Port of Spain denied him this job, but, in 1852, Stollmeyer took over editorship of the *Trinidadian* weekly newspaper. Its previous editor, George Numa Dessources, had departed with a hundred settlers to found another (failed) encampment in Venezuela, called Numancia. "Never despair," Stollmeyer exhorted these readers, seemingly unaware of his own checkered reputation. "You will have to suffer and work much and hard in the beginning to obtain 'PARADISE WITHOUT LABOUR' in the end."[22] Stollmeyer's calculus was shifting: leisure would now arrive after an investment of sweat equity. On Trinidad, too, the German now recommended felling, hoeing, and weeding. "This fertility of the soil," he reported in late 1852, "in a very short time repays the toils of the industrious laborer."[23] The editor was responding to an alarming development. Large numbers of ex-slaves had exited the countryside, regarding agriculture altogether as oppressive (Wood 1968, 48). Colonial legislation restricted their rise in urban professions. The frustrated black man, Stollmeyer wrote, "leads a careless life, and falls into indolence and dissipation, . . . branded with the epithet of the 'lazy, worthless, slovenly, dirty creole.'"[24] Did Stollmeyer himself subscribe to this antileisure thinking? Certainly, his argument cast labor positively—as a good withheld by greedy whites. Stollmeyer still believed in fecundity—now complemented with toil. The editor of the *Trinidadian* began to embrace the agrarian muscle that he had come to Trinidad to abolish.

In the next year, 1853, Stollmeyer's interests expanded from farming to mining and manufacturing. Focusing on pitch, he praised Cochrane for "the establishment of this new branch of industry [that] will become the most important event in the history of our island."[25] "May we not then cherish the hope," Stollmeyer asked his readers, "that from this accumulation of wealth a radical change will take place throughout the civilized world towards the amelioration of suffering humanity . . . ?" Yes, answered the editor, but he no longer equated "suffering" with work. Quite the opposite: the same editorial referred to "thousands and thousands who will be employed in the different manufactures to which our mineral treasure is susceptible."[26] He proposed a new settlement scheme, this one importing freed blacks from the United States. Stollmeyer promised them "a home . . . where in a short time they may be able by their industry, and assisted by the productiveness of our soil, to acquire a position for themselves." And, further, "with a population reinforced by thousands of industrious immigrants . . . provisions [food crops] would become abundant."[27] Here, Stollmeyer married this new value of industry and hard work to a productive, tropical biome. A few years later, he discovered a conflict between these two forces, wherein nature undercut industry. "This climate and the rich soil of the island," he wrote in a private letter in 1855, "enable men who have very few wants to live isolated as squatters idling away 9/10 of their time." They lived like maroons, provoking a moral unease similar to that of Chacón. This time, blacks in the hills posed no military threat, but Stollmeyer echoed the former governor's more diffuse concern: lazy freedmen, he wrote, "have only the animal instincts developed."[28] In short, Stollmeyer reversed the moral positions of toil and leisure. What was once virtue turned to vice, and the paradise without labor became a contradiction in terms. Energy lost its conscience.

A second public humiliation probably accelerated this ideological conversion. This scandal unfolded in early August 1853 as a certain Mr. J. Kavanaugh barred Stollmeyer from entry to his private club. The disagreement centered on Stollmeyer's nonpayment of membership dues, on Kavanaugh's objection to one of Stollmeyer's editorials, or on both of these issues. At any rate, Stollmeyer publicized the conflict in his newspaper. Feeling himself libeled, the elderly Kavanaugh, shortly thereafter, hit Stollmeyer with a stick in the streets of Port of Spain. The next issue of the *Trinidadian* decried this "brutal assault," which very nearly rendered "our beloved wife and our five children orphans."[29] Stollmeyer pressed charges and demanded damages of 500 pounds. The judge found Kavanaugh guilty of assault but cut his damages to 5 pounds. The defense debunked everything from the thickness of the stick to the ferocity of his seventy-year-old assailant to Stollmeyer's self-described counterattack reminiscent of "our dueling days." Finally, the defense criticized his grammar. Referring to him sarcastically as "Sir Oracle," the attorney asked, "Was there ever anything so ridiculous as the editorial WE in this case?"[30] Stollmeyer reaffirmed, "We have taken up the pen in the holy warfare of the Press against oppression, ignorance, vice, and intemperance."[31] By the end of it, though, he may have been looking for a way to leave the *Trinidadian*. His final editorial mentioned the destruction of the paper's printing press through fire but characteristically gave greater weight to his own shift in preferences. "It is time for us to think about more profitable modes of employing our time," he wrote, still with the royal pronoun, "and to concentrate all our efforts upon the fashionable and praiseworthy object of making money."[32] Greed replaced generosity as black goo oozed from the earth.

In fact, pitch and other hydrocarbon fuels proliferated in the 1850s, and Stollmeyer made a careful, if utterly mercenary, choice among them. The sugar refineries burned wood and crop residues known as megasse. These cane stalks constituted a diffuse, rather than dense, fuel. To keep cane juice boiling, stokers had to transfer megasse rapidly and continuously (Mathieson 1926, 64; Mintz 1985, 50; Smil 1994, 117). Bending constantly and under intense heat, they worked—the social critic Lewis Mumford (1934, 235) later noted—as virtual "galley slaves" long after emancipation. By the 1850s, Trinidadian growers had mostly shifted to coal imported from

Britain (de Verteuil 1848, 77). The new fuel probably improved working conditions in the factories: since coal's energy density stands at double that of crop residues, coal could easily cut the stoker's lifting job in half. Bitumen, which was available locally, might achieve the same efficiency. Stollmeyer advertised this hope as fact in 1853. The *Trinidadian* ran his signed advertisements for "cheap fuel . . . equal to coals," cash preferred.[33] By 1871, experiments sponsored by Stollmeyer and Cochrane proved the energetic equivalency. "With perfect combustion," Stollmeyer touted to planters, "a ton of raw asphalt will give as much heat as a ton of the best of the best stone [anthracite] coals." The new, island-produced hydrocarbon, then, promised to make the labor savings achieved by coal economically sustainable in the long term. The same publicity, however, suggested an additional step, cutting the asphalt into smaller pieces. A machine would actually split the asphalt, but "manual labour" would sift and transfer the pieces.[34] Still, shifting from megasse to bitumen—or even adding bitumen to the fuel mix, as occurred more frequently—surely alleviated some of the toil in sugar factories. If so, the ex-utopian Stollmeyer reduced work only as an unintended by-product of business objectives.

Indeed, Stollmeyer rejected a second petrolic means of saving labor as simply unprofitable. Also in the pivotal 1850s, he and Cochrane perfected a technique for distilling pitch into kerosene. Their Trinidad Petroleum Company produced the fuel for illumination in Port of Spain and further afield (Wiltshire 2007, 23). Why did Stollmeyer and Cochrane not send kerosene to the sugar factories? Kerosene's energy density exceeded that of bitumen and anthracite coal by a significant margin (figures 2.2 and 2.3). At the boiler, a stoker would have to lift and move proportionately less kerosene than coal. Furthermore, the liquid quality of kerosene might obviate stoking altogether. Kerosene could travel in pipes, either with gravity or with relatively easy pumping uphill. The fuel could convert at least one stage of sugar production into a paradise without labor. Yet no country besides tsarist Russia—which produced enormous volumes of kerosene—replaced coal with kerosene (or petroleum) in the nineteenth century (Mitchell 2011, 31).[35] Elsewhere, kerosene did what only kerosene and whale oil could do: keep the lights on. And, beginning in the 1870s, Trinidad's whalers—along with the entire Atlantic fleet—exhausted the supply of whale oil (de Verteuil 2002, 254–55). Cetacean extinctions forced the price of kerosene up. From then on, one might sell kerosene for pur-

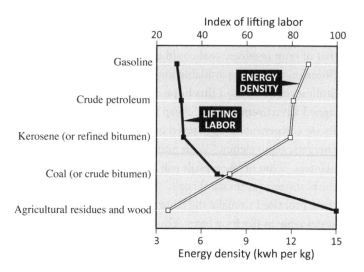

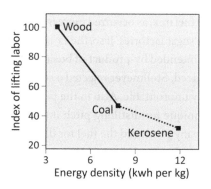

2.2 Energy density and lifting labor associated with various fuels. Prepared by Mike Siegel of Rutgers Cartography Lab.

2.3 Energy options for Trinidad's sugar boilers. Prepared by Mike Siegel of Rutgers Cartography Lab.

poses other than illumination, but only at an opportunity cost. Stollmeyer obeyed these incentives of supply and demand almost unthinkingly, missing multiple chances to alleviate toil. In 1866, not far from the Pitch Lake, Walter Darwent had drilled one of the first oil wells in the world. Petroleum, which was also distilled into kerosene, quickly undercut Stollmeyer's share of the illuminant market. Stollmeyer might have then developed other markets for kerosene, where it would have replaced somatic power. Instead, he gambled upon the exhaustion of oil. "It would be safer to speculate," he wrote in 1866, "upon the great prize in the Frankfurt lottery than upon the finding of oil-wells in Trinidad."[36] He gambled wrong, of course, but only money—not liberation—was now at stake.

With this change of heart—perhaps accelerating it—Stollmeyer's latent bias came into the open. The sight of freedmen enjoying their leisure actually appalled him. He confided these sentiments to relatives and friends, rather than to readers of the *Trinidadian*. In 1854, he complained to his mother, "The former conditions under slavery have spoilt the labourers here and they hate to work."[37] "The Blacks are as lazy as possible," he wrote to an American utopian in 1855, "ignorant rum drinkers with few exceptions, they are worse off than in the time of slavery."[38] His complaint echoed—and may have borrowed from—the famous 1849 tract of Thomas Carlyle. Published in London, his "Occasional Discourse on the Negro Question" referred to freedmen as "indolent two-legged cattle . . . 'happy' over their abundant pumpkins" (Carlyle 1849, 675; cf. Wahab 2010, 167–68). In Trinidad itself, however, such reactionary sentiments soon became unfashionable. In 1858, Louis de Verteuil's definitive geography of Trinidad both documented and excused laziness: "The slave [was] obliged to toil for the benefit of his master," explained de Verteuil, and "viewed the obligation of working as a curse" (1858, 489). Cochrane expressed even greater sympathy. Caribbean blacks, he opined, were willing to work for "those [rare planters] who pay a reasonable price for work and are punctual in the payments" (Dundonald 1851, 109–10; cf. Lloyd 1947, 201). How could one business partner detect eagerness where the other saw only sloth? Perhaps Stollmeyer agreed with Etzler, who, back in 1833, had advised readers in the United States: "The slaves in your country will cease to be slaves. . . . New mechanical means will supersede their employment. . . . You may then easily dispose of this unfortunate race. . . . Send them to some distant part of the world . . . and make amends for the grievous wrongs they have suffered in this country" (Etzler 1833, 14). At root, these utopians did not devise machines so as to liberate slaves from their masters. They hoped for the reverse: to liberate masters from their slaves—and expel these redundant humans.

Trinidad conceived many possibilities regarding the ability to do work. Some of them, however, emerged stillborn. Among those failures, oil missed its labor-saving potential due to a certain convergence of intolerance and greed. Although abolitionist, Stollmeyer developed a sharp prejudice: the entrepreneur's resentment of loafers. By his own admission, Trinidad's freedmen were living in conditions analogous to Fourier's Phalanx. They enjoyed their leisure and helped each other to get by. Absent the iron slave, tropical forest could still generate sufficient nutrition under minimal cultivation. Even today, Trinis collect mangoes growing wild or untended on public or private land—a practice criminalized as "praedial larceny." Before 1850 or so, Stollmeyer would have admired this carefree, unstraining harvest. After that point, however, the German became one of those "interested parties . . . preaching against idleness." He built his business with a strict Protestant work ethic, expecting the same of others. The extraction of bitumen, in fact, required much sweat and striving. Men hacked at the Pitch Lake with pickaxes and hoisted the material into handcarts. Only distillation—later in the commodity chain—produced the liquid fuel that flowed and could replace labor power. Stollmeyer appreciated the former quality to the exclusion of the latter. "The Pitch Lake is a mine of wealth," he wrote to an investor in 1871, "and, if properly, judiciously, and liberally handled, can make fortunes for all at present concerned in it."[39] Stollmeyer made his own fortune partly, it was alleged, by cheating workers in La Brea through the company store (Massé 1988, 298). In his later years, he directed Trinidad's electricity, telephone, and ice manufacturing companies (Wood 1968, 89). None of these businesses would have pleased the earlier Stollmeyer, the man who named his son Charles Fourier. In 1904, Stollmeyer's obituary praised him for precisely the values he had once sought to render obsolete: "industry, energy, [and] indomitable perseverance."[40]

Beyond this one man, oil affected the balance of work and rest in ways that were contingent and unpredictable. One U.S. gallon of crude oil contains the labor equivalent of nearly six hundred person-days.[41] As this potential came to light, hydrocarbons enhanced the productivity of human labor. Turbocharged in this way, a worker could finish the day's quota shortly after breakfast. Few have ever appreciated the radical possibilities at hand. In 1880, the Cuban-born socialist Paul Lafargue concluded his essay titled *The Right to Laziness* with a paean to coal power: "the machine is the redemption of humanity, the God that will redeem the man from *sordidæ artes* and wage work, the God that will give him leisure and liberty."[42] Lafargue expected what one might call a rest dividend. This "right to laziness," however, never rose above the legal status of an undeserved privilege. Colonial governments disparaged it, elaborating a "myth of the lazy native" who refused to work for whites (Alatas 1977). As if to disprove the insult, anti- and postcolonial manifestos and constitutions disavow any right of repose. Such documents are more likely to enshrine a right to work. In so doing, they exploit the opposite side of the hydrocarbon equation: a worker who finishes the first task in short order may finish many more by closing time. Hydrocarbons allowed workers to do more. Fossil fuels, in other words, enabled and still enable either more rest or more activity. The former option appealed to Fourier and Lafargue—but not, by and large, to states and corporations. Backed by these interests, hydrocarbons have driven more activity and the frenzy of capitalist growth. There has been no energy transition, only the layering of fossil on somatic power.

As a symbol, on the other hand, oil promoted a subtle transition from one form of morality to another. Or, rather, the amorality of oil replaced energy systems saturated with religious and ethical meaning. Gumilla fairly worshiped the attractive virtue of solar rays. Slavery and wage slavery always offended the conscience of many people, not least among those wearing the shackles. Participants in the trade, such as Chacón, could not avoid the awkwardness of treating people like fuel. Petroleum, by contrast, provoked neither offense nor unease. Since it did not spread leisure, the new fuel hardly incited celebration or romance either. Petroleum made drillers and refiners wealthy, but it did not associate them or itself with that signal value of the Caribbean, freedom. Oil, in a word, did not rise above the ordinary. Again, though, different contingencies might have produced different outcomes, as they have in other cases. Diamonds imply permanence, fidelity, and a love that will not weather. Fabricated by ad men, this ideal topples easily. As the blood diamond campaign traces gems to sites of civil conflict and slave labor, diamonds increasingly imply devastation and death. Luxury flips to exploitation as one glances from right to left in Amnesty International's poster (figure 2.4). Cigarettes have undergone a similar inversion of sentiment. Until perhaps the 1970s, smokers associated tobacco with sophistication, elegance, and even health. Medicine shattered this illusion. Now, especially in the United States, cigarettes increasingly connote ignorance, incompetence, and disease. Admirers of a commodity, then, seem prone to reevaluate it. Extreme romance may flip to extreme antipathy, as if the commodity itself betrayed a sacred trust. No such about-face has occurred with respect to oil, and cultural conditions do not favor one. There is no trust, admiration, or romance to betray. Stollmeyer, like oilmen elsewhere, declined to invest hydrocarbons with

Quel prix pour ces diamants ?

Non au commerce des armes et des matières premières avec les pays qui violent les droits humains. Amnesty international
www.amnesty.asso.fr

2.4 *For What Price, These Diamonds,* from 2003 billboard campaign "Stop arms and raw materials trade with countries that violate human rights." Courtesy of the French Chapter of Amnesty International (AISF).

such grandiosity. Oil is always already a cynical category. One might almost believe that, after the emotional roller coaster of slavery and emancipation, the designers of energy sought a source of terawatts without drama, without a "commodity affect" (Mankekar 2004, 408). Intentionally or not, they got that: a fuel so flat that the protester finds little traction with which to advance. Except when oil spills locally, one treats it as a means, an instrument toward the things that really matter. Useful as it is—perhaps, like money—oil only rarely touches questions of moral worth. Banally and too easily, hydrocarbons flow and spill everywhere.

PART II

ORDINARY OIL

Author's son, Jesse, with oil pump jack, Point Fortin, Trinidad, 2010

SILENCES ARE NOT ALWAYS QUIET. They can resound with noise and quite articulate speech. They are, as I wrote in the introduction, the absences that present a shape. This second half of *Energy without Conscience* explores the contours and discourses surrounding—and, in a sense, obstructing—what is for me the core issue: a moral reckoning with hydrocarbons and a sense of responsibility for climate change. Oil appears all too banal and ordinary. In this more ethnographic section, my informants grapple with notions of plenty (chapter 3), with industrial accidents in their neighborhoods (chapter 4), and with environmental victimhood (chapter 5). Expert and popular opinions proliferate in what seems like a robust debate. Yet nearly all participants draw back from the cliff's edge. They refuse to consider questions of conscience: if (rather than how fast) one should produce oil, or whether oil is intrinsically (not incidentally) harmful, or whether they have perpetrated (not merely suffered from) climate change. In my fieldwork, a handful of self-aware geologists and policy makers appreciated these dilemmas. Far more often, their own expertise and activism proved so interesting that it distracted them from considering alternatives. Perhaps—if silences are loud—then complicity is diverting and fulfilling in this way. My informants did not cover up a shameful secret, as one might imagine knowing perpetrators of harm to do. Climate change, they understood, was important, and they would deal with it. But they always found something more pressing: oil to locate, toxins to fight, or worse offenders to indict. This is the most widespread, least reproachable form of complicity: an earnest pursuit of local, immediate, rather ordinary concerns in the run up to apocalypse. Like stewards rearranging deck chairs on the *Titanic*, one can easily lose a sense of proportion.

But who I am to criticize these well-meaning Trinidadians? Many of them, after all, cope with economic circumstances far more adverse than those of a university professor (although the energy executives enjoy far better conditions). Before such subalterns, ethnographers usually defer. Waiter-like in their humility, they act as if the customer-informant is always right (Rabinow 1977, 45). My informants, I concluded, are mostly wrong—either mistaken on ecological grounds or conducting environmental malfeasance. And I write forthrightly in that conviction not only because it is true but also because it matters to us all. Here again, one might ask why I make Trinidadians' affairs my business. The Indian social critic Vandana Shiva famously accuses North Atlantic environmentalists of practicing an imperialist "global reach" when they insist, say, that African peasants refrain from hunting animals. I agree with her in that instance. Hydrocarbons are different. More so than any other form of environmental harm or violence, they circulate through the biosphere. Natural gas burned in or exported from Trinidad circumscribes lives elsewhere. Coastal residents of Bangladesh or Vietnam have perhaps the greatest cause for concern. Still, Superstorm Sandy—which hit New Jersey after the bulk of my fieldwork—made the threat to me, my family, and my community apparent. The Trinidadian energy companies I study bring danger to my doorstep. They are an empire. So I write with as much anticolonial outrage as colonial arrogance. But above all—and to put aside ill-fitting metaphors—I try in this ethnographic section to capture the frustration and possibility of my own encounter with climate-changing complicity.

CHAPTER 3

The Myth of Inevitability

Picture a gusher. Microorganisms have photosynthesized solar radiation. They and the zooplankton who preyed on them have died and fallen to anoxic depths. Then, over scores of millions of years, sediment has buried this organic matter and subjected it to pressure and heat, baking it into hydrocarbons. Eventually, pressure has forced the substance upward again and, ultimately, into the drillable strata, from whence it may gush. That black fountain crystallizes all the economic, social, and psychological dramas of petroleum into one vivid image (Ziser 2011, 321). Recall, if you have seen it, *Giant*, the 1956 epic film of the Texas oil patch. Petroleum rains down sensually on James Dean, transforming him from a lowlife punk into the crassest tycoon. Today, blowout preventers almost always throttle such spills. Still, the gusher maintains a presence in the assumptions, terms, and measurements of oil firms. Experts describe the commodity's path as a "stream." Hydrocarbons extracted upstream descend to midstream refining and downstream finishing as plastics or fuels. Gravity—one might imagine—pulls oil to its destination. The industry does not even refer to its business as "extraction." "That is what mining companies do," explained one executive in Trinidad, shocked that I would confuse his business with such a nasty affair. An offshore platform, he explained, "produces" oil in the way one might produce a fork from the kitchen drawer. The company brings forth materials that seemingly belong on the planet's surface. One doesn't need—in the fashion of miners—to go down there personally and haul them up. Another term illustrates this sense of modest effort. An oil and gas company "recovers" hydrocarbons, if they are "recoverable," or employs "secondary recovery" to acquire stocks previously considered "unrecoverable." Oil virtually produces itself. Most of it seeps around caprocks—tellingly denoted as "traps"—and rises to the surface. The same Trinidadian expert assured me that 95 percent of it had come up naturally over geological time. He felt duty-bound to bring home the rest. Otherwise—to use the final bit of lingo—hydrocarbons currently impossible to produce remain "stranded" (Bridge 2004, 396). Oil and gas, in short, come up alone, ascend in a rescue operation, or await a delayed release, which is both right and inevitable. Where is the space for deliberation—for conscience—in this seemingly natural plot?

Skeptics do pose two sorts of challenges, a constraint related to supply and another related to pollution. According to the first objection, stocks are limited. Organic matter must bake under pressure and natural radiation for millions of years to become bitumen or asphalt, longer to become petroleum, and longer still to become gas. At a very slow rate of consumption, hydrocarbons are renewable. But industrial societies have exceeded that geological pace by orders of magnitude. Everyone in the industry understands this fact. Only a handful of amateur geologists believe in rapid oil generation, and they clamor ineffectually from the sidelines. Still, this ultimate limit on supply is easy to ignore. It seems too far out—rather like the way in which the sun's finite nuclear fuel limits agriculture. Those concerned about oil supplies notice a proxy variable for dwindling supplies: effort and expense. As companies remove oil from the earth's crust, it becomes harder and harder to find more. In this view, ultradeep drilling, hydrofracking, and tar sands all signal physical and financial strain and the beginning of the end.[1] Are the 1962 predictions of M. King Hubbert coming true? Hubbert anticipated a moment when the rate of extracting hydrocarbons would fall below the rate of locating new sources. The United States passed this point of "peak oil" in the 1970s.[2] The world may have already passed it. For various reasons—particularly centering on unaudited logs of Saudi Arabia's huge Ghawar Field—peak oil seems to be unverifiable (Simmons 2005). Past the peak, oil producers and consumers will descend into a "long emergency" of squabbling over scarce, expensive hydrocarbons. Amid wars, life becomes nasty, brutish, and short (Kunstler 2005; Roberts 2004).

The second limit on production raises a similar specter of decline and strife. This constraint—which I call the "climate boundary"—also depends on a mismatch of rates. Oceans, forests, and soils can fix in biomass and otherwise neutralize a certain amount of atmospheric carbon dioxide per year. Humans could burn hydrocarbons at a stately, sustainable pace.

Current combustion, however, substantially exceeds that rate, threatening to overwhelm stabilizing mechanisms and throw the climate into accelerating shifts. In 2007, the Intergovernmental Panel on Climate Change established a warming of 2°C, leveling off in 2050, as the maximum tolerable level. By tolerable, they meant something less than comfortable: atoll states would disappear. Warming beyond 2°C would inundate the far more populated coastlines of South and Southeast Asia. How much oil can industrial societies burn before provoking such human misery and death? In a 2012 article titled "Global Warming's Terrifying New Math," the activist Bill McKibben compared the carbon content of the world's proven reserves fossil fuels with the climate boundary. "We have five times more oil and coal and gas on the books," McKibben concludes, "[than] the climate scientists think is safe to burn."[3] The world will run out of storage space for burned hydrocarbons long before it runs out of hydrocarbons themselves. The sink, in other words, limits industry more severely than does the supply. Again, this diagnosis is not widely disputed—except by deniers of climate change itself. Within energy firms, their influence is both waning and redundant. The industry has found other ways to insist upon the banal inevitability of oil despite and beyond the threshold of safety.

Those arguments surround the silence of complicity. Myths of oil circumscribe, delimit, and obscure the moral reckoning with hydrocarbons and climate change. They always found something more pressing: oil to be located. In large part, diagrams made this choice of priorities, and the evacuation of conscience, appear normal. Geologists picture the underground in ways that suggest permanent, inevitable flows. This simplified view of the world resembles other notions of capitalist growth and technological advance: it depends upon particular ways of seeing.[4] Among Trinidad's oil experts, I investigated the charts and their dissemination. For a century and half, geology had emphasized the dynamic quality of the earth's crust. Substances at depth rose—if not in the past, then now, and if not now, then later. In the lifetime of my informants, geologists had married this model of upward flows to the economic scenarios of supply and demand. Hydrocarbons, they had come to assume, left the ground and entered the global market in one natural, entirely ordinary progression. To provoke conscience, I occasionally cornered people in conversation. "Fuck you, no apologies, oil is here to stay," they might have shouted, as ExxonMobil apparently did to one observer.[5] Fortunately, my West Indian informants spoke more politely and, indeed, cared about climate change. Yet even their efforts to reduce carbon emissions actually produced more oil—and more carbon emissions. It could not be otherwise, they assured me.

Making the Y-Axis

Geology is a science of vertical movement. The things that move are huge and heavy and move very slowly. So says the uniformitarian theory, published in 1830 by the Scotsman Charles Lyell. In some ways, such gradualism defies belief more than did the earlier catastrophist notions of rapid, biblical creation, flooding, and so on. To follow the vertical movement of continents, one must inhabit what the environmental writer John McPhee (1980) calls "deep time." Over millions of years, eroded sediments may turn a floodplain into a plateau. One plate will dive down, deflecting its adjacent plate upward. Describing these acts of elevation requires an otherworldly lexicon: Holocene, Pleistocene, Pliocene, Miocene, Oligocene, Eocene, and so on going back to Earth's pre-Cambrian beginning as a lifeless, cooling moon of the sun. The communicative art of geology lies in making this deep time comprehensible without, at the same time, utterly dispelling its strangeness. Verticality helps strike that balance. The amateur may more easily grasp a thousand feet than a million years. The up-and-down axis, in fact, compresses time. Then, superposition—the principle that layers fall sequentially upon one another—translates descending distance into antiquity and shallowness into newness. Geologists distinguish strata as upper and lower and periods as late and early, but, as often as not, they interchange the terms. The past stretches as an arrow piercing the heart of the earth. One need not imagine much to extend that arrow into the future—as thrusting, gushing, and seeping movements up through and out of the crust. The pressures of profit can easily bake geology into such a predictive belief. In this way, Trinidad and other oil crucibles produced what one might call a vernacular science of hydrocarbon uplift. As in many technical fields, petro experience affords little space for alternatives and less for the most challenging ethical questions.

That amoral vernacular is more visual than linguistic. It relies upon images of deep time, in whose invention the small island of Trinidad played an outsized role. This "visual language for geological science" emerged in the course of the nineteenth-century uniformitarian revolution (Rudwick 1976). At midcentury, Britain's Geological Survey applied that theory to "see geologically" in Canada and elsewhere in the empire (Braun 2000, 22; Stafford 1990). To Trinidad, the survey sent geologist George Wall, accompanied by the artist Jas Sawkins. Their 1860 report made use of two newly available diagrammatic forms: the cliff face and the traverse section.[6] In the first of these techniques, Sawkins simply set his sketch pad on the beach— probably close to present-day Radix on Trinidad's east coast—and reproduced the strata he saw. Although he surely overemphasized the boundaries between rock types, his drawing reproduced the original proportions of strata, scaled against the human form at water's edge (figure 3.1). The cross section, by contrast, took more liberties with the landscape. Consider Sawkins's section of the Pitch Lake (figure 3.2). Neither he nor Wall ever saw any such sandwich of shale, sand, and shale. Rather, as Rudwick (1976, 164) writes, the cross section conducts a "thought-experiment," proposing how the landscape might appear if sliced vertically. Note the absence of any scale in the section, an omission that suggests this hypothetical quality. The geological report did not include a columnar section, the third type of vernacular image. Rather than depicting a particular place, this kind of cross section aggregates and interpolates geological layers. In 1912, Edward Hubert Cunningham-Craig published one of the earliest such diagrams related to Trinidad. This British early petro-geologist studied the Pitch Lake as well as similar outcrops in the nascent oilfields of Burma and Persia. Pictographically, he inserted a "Cretaceous inlier" (from 144–66 million years ago) beneath younger sediments bearing "manjak," a local bitumen variety (figure 3.3). From top to bottom, in other words, the cross section descends in reverse chronological order. Time ran parallel with depth.

During the first three quarters of the twentieth century, British and American scientists both found more oil in Trinidad and drew increasingly refined sections of it. Especially in the heyday of the 1950s, detailed traverse sections represented the high art of exploration. They converted abstract space into known places (Carter 1987, xxiii). A 1958 rendition of the East Penal field—by Peter Bitterli of Shell—catches the eye more

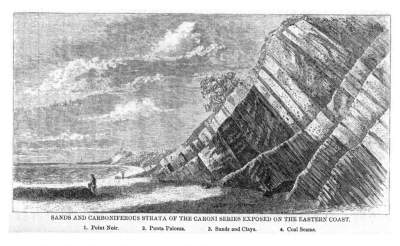

SANDS AND CARBONIFEROUS STRATA OF THE CARONI SERIES EXPOSED ON THE EASTERN COAST.

1. Point Noir.   2. Punta Paloma.   3. Sands and Clays.   4. Coal Seams.

3.1 Wall and Sawkins's diagram of a cliff face, 1860.

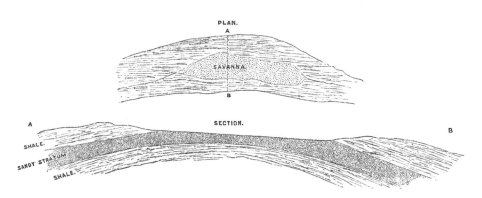

PLAN.

SAVANNA

SECTION.

SHALE.

SANDY STRATUM

SHALE.

3.2 Wall and Sawkins's traverse section of the Pitch Lake, 1860.

than any other (figure 3.4). As depicted, an unconformity of overthrusting Eocene shale traps petroleum in Herrera sands. Then, descending from the upper left, oil wells puncture that caprock and release the hydrocarbons. This and other columnar sections guided the drill bit in a conceptual fashion: charts created a geological and economic abstraction. Let's consider a drawing first published in 1958 by a threesome of geologists at the major firms on the island. Titled "Summarized Miocene Stratigraphy of Southern Trinidad," the graph represents rock formations as horizontal layers associated with prehistoric eras. Columns refer to regions of Trinidad, from southwest to southeast (figure 3.5). "Miocene stratigraphy," the

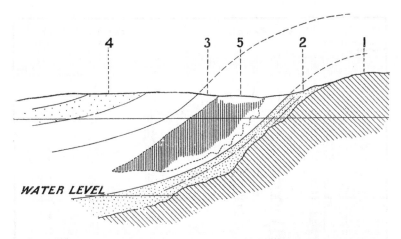

WATER LEVEL

Fɪɢ. 8.—Diagram illustrating mode of occurrence of Manjak veins in Trinidad (San Fernando Field). 1. Cretaceous inlier; 2. Oil-bearing sand; 3. Clay; 4. Sandstone; 5. Zone containing Manjak veins.

3.3 Cunningham-Craig's columnar section, 1912.

authors indicate, "is essentially a clay-silt-sand sequence which is divided into three major sedimentary cycles, separated by local unconformities" (Barr, Wait, and Wilson 1958, 536). Those unconformities—which might twist and turn bizarrely—appear nowhere in the summarized stratigraphy. Summary required excluding them. Indeed, it required bringing all the data into conformity with a linear time-depth axis. Barr and his colleagues flattened the undulating underground topography to show the past simply as depth. The y-axis was time, and—when implied in traverse sections—the vectors pointed up. Taken together, these works of 1958 show oil in ancient layers and oil rising more recently through layers. Petroleum geology thus mastered the art of describing and depicting upward migration.

In Trinidad and Tobago, no contemporary geologist has represented this verticality with greater expertise or enthusiasm than Krishna Persad. Following independence, his generation took over from the geologists of British and American firms. Texaco, in fact, gave Persad a scholarship to study chemistry in the 1960s. Midway through his doctorate, he switched to geology, and this combination of disciplines has arguably made him the island's most successful independent oil producer. Not merely a businessman, Persad founded the Geological Society of Trinidad and Tobago.

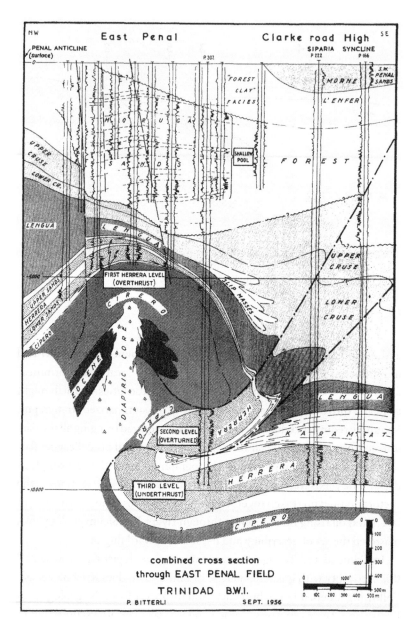

3.4 Bitterli's combined cross section, 1958. AAPG © 1958. Reprinted with permission of the American Association of Petroleum Geologists (AAPG), whose permission is required for further use.

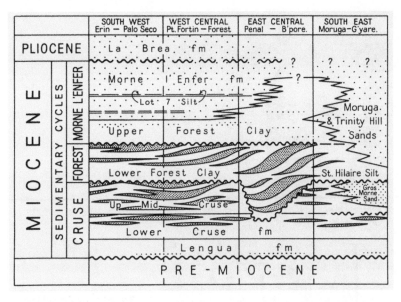

			SOUTH WEST Erin — Palo Seco	WEST CENTRAL Pt. Fortin — Forest	EAST CENTRAL Penal — B'pore.	SOUTH EAST Moruga-G'yare.
PLIOCENE			·La· ·Brea· ·fm·			? ? ?
MIOCENE	SEDIMENTARY CYCLES	MORNE L'ENFER	Morne · · l'Enfer· ·fm		?	Moruga.
			Lot· 7 ·Silt			& Trinity Hill
		FOREST	Upper	Forest	Clay	Sands
			Lower Forest Clay			St. Hilaire Silt
		CRUSE	Up—Mid. Cruse			Gros Morne Sand
			Lower Cruse		fm	
			Lengua		fm	
			PRE - MIOCENE			

3.5 Barr, Wait, and Wilson's summarized stratigraphy of southern Trinidad, 1958. AAPG © 1958. From the American Association of Petroleum Geologists (AAPG), whose permission is required for further use.

(Virtually all Trini geologists practice petroleum geology.) In his spare time, he travels, paints, and draws (figure 3.6). In 1993, Persad and his wife published *The Petroleum Encyclopedia of Trinidad and Tobago*. The volume reassures readers: "There is little cause to worry [about declining oil supplies] and . . . in fact, we can expect oil production to continue at significant levels for decades to come" (Persad and Persad 1993, 4). In the next two decades, even as Trinidad's oil output fell nearly to zero, Persad compiled the magnum opus proving that it need not be so. In 2011, he completed *The Petroleum Geology and Geochemistry of Trinidad and Tobago*. The volume interprets strata according to plate tectonics—which the geologists of the 1950s had not accepted—and in light of the chemical principle of evaporative fractionation. The charts themselves reproduce earlier work, going back, in fact, to the 1950s. But they also innovate in one key respect: they show the migration of oil. Oil, it is now known, matures from deposited organic matter in formations of between 7,000 and 18,000 feet in depth. This layer is known as the "oil window" (Hyne 1995, 171). From that source rock, petroleum moves upward and laterally, pushed

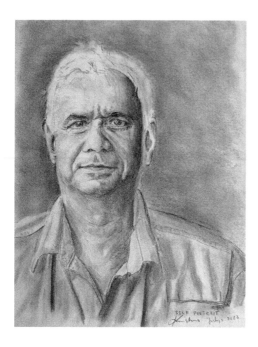

3.6 Krishna Persad (self-portrait), 2012.

by underground pressure and by gravitational sorting with water. (Oil is lighter than water.) Trinidad, Persad argues, has experienced multiple bursts of tectonic movement and faulting, leading to "oil migration pulses." A characteristic sketch shows petroleum rising along diagonal pathways marked with arrows (left of figure 3.7). Like Bitterli, Persad focused on the unconformities. At the same time, two additional lines recall Barr's summarized stratigraphy: the horizontal line marked "top [of the] oil window" and the vertical arrow labeled "speculative prior . . . pulse." With greater certainty than ever before, Persad's chart gives oil a position and an upward pathway from it. Oil would flow until stopped by a trap. Then, to raise production, oil companies would need only to perforate such rock at the points Persad indicated.

This sense of possibility and optimism brightened my reunion with Persad shortly after the publication of *Petroleum Geology and Geochemistry*. (We knew each other already.) "This is my legacy work," he said of the new book, agreeing to a discounted price that reflected the lean state of public higher education. We talked about oil fractions and the superposition of gas and lighter oil above heavy, waxy crudes. Trinidad was exhaust-

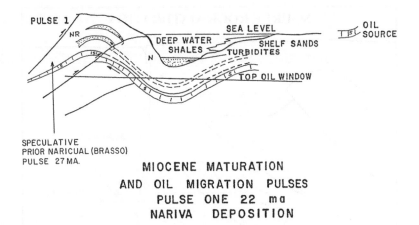

MIOCENE MATURATION
AND OIL MIGRATION PULSES
PULSE ONE 22 ma
NARIVA DEPOSITION

3.7 Persad's oil migration pulses, 2011.

ing the former, but their very existence indicated that dense petroleum still lay deeper. "You could be looking at double, triple the reserves," he exclaimed, as we sat in his home office in South Trinidad.[7] Possessed of a gentle, good-natured humor, Persad frequently slipped between dense science and less scientific buoyancy. We met again the next month—February 2012—in Port of Spain at the Energy Chamber's annual Energy Conference. To the chamber's large membership, Persad presented "Finding Oil in T&T's Unexplored Acreage." The acreage lay below explored strata. "If you drilled deep," he almost pleaded to the suit-attired executives, "you would find black oil." The oil itself was going half the distance. Persad's second slide, titled "Source Rock Maturation," united two concepts usually considered separately: the generation, or maturation, of oil in deep sediments and its later migration into shallower formations (figure 3.8). In the cross section, one pathway took oil all the way to a surface seep. The slide show ended with a note of hope and a surprisingly precise scenario: short-term, medium-term, and longer-term estimates culminated in a "total potential upwards of 3 billion bbls [barrels] recoverable."[8] The geologist's eyes twinkled at the audience. Such confidence, of course, flowed like the liquor at the Energy Conference. Persad enjoyed himself amid the glitter of the Hyatt Hotel. Having once told me any conference was worth going to "if it helps me produce a barrel of oil," he made more petroleum seem guaranteed.

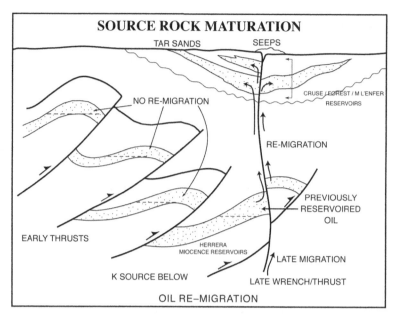

SOURCE ROCK MATURATION

TAR SANDS   SEEPS

NO RE-MIGRATION

CRUSE / EØREST / M L'ENFER
RESERVOIRS

RE-MIGRATION

PREVIOUSLY
RESERVOIRED
OIL

EARLY THRUSTS

HERRERA
MIOCENCE RESERVOIRS

LATE MIGRATION

K SOURCE BELOW

LATE WRENCH/THRUST

OIL RE–MIGRATION

3.8 Persad's migration and maturation of oil, 2012. Prepared by Mike Siegel of Rutgers Cartography Lab.

Proving Up

Abundance is not always desirable. Purveyors of goods frequently wish to represent their supplies as scarce, especially to those who might question a high price. Downstream from the wells, manufacturers of plastics or fertilizer are constantly trying to acquire hydrocarbons more cheaply. So too is the car driver, watching numbers dial upward at the gas pump. To these buyers, the oil industry invariably asserts that its commodity is scarce and difficult to obtain. Almost stubbornly, however, oil continually proves to be available. Since at least the 1920s, while producing more and more oil, firms have consciously confronted a problem of surfeit. As they extract oil, they must also "produce scarcity." Or, as Gavin Bridge and Andrew Wood surmise, executives ask themselves, "How [can we] organize scarcity in the face of prodigious abundance" (2010, 565–56)? So-called aboveground risks—such as violence and boycotts—help manufacture scarcity. The Arab Oil Embargo of 1973, two wars against Iraq, and ongoing sanctions against Iran have periodically tightened global hydrocarbon markets artificially. Yet no producer wants altogether to disguise plenty. An industry running out of raw materials has no future. To investors, therefore, oil firms must represent oil as prolific, such that a billion dollars spent on drilling for oil will find oil. The problem, then, centers less on producing scarcity than on telling one story to consumers and another story to investors (Olien and Olien 2000, 137; Chapman 2013, 97). The spokespeople of petroleum must keep their audiences separate and avoid confusing their half-truths. In the 1970s, international bodies devised a set of terms and diagrams that bifurcated their message graphically—into a story about large resources and small reserves. Oil "proved up" from one to the other. That narrative has mostly stuck, persuading the people who need persuading. It unraveled in Trinidad, however, in the late 2000s. There, in a small country, investors and consumers sat in the same room—often as the very same people—and pushed that contradictory message to its limits. Still, they never pushed it far enough to doubt the rightness altogether of oil and gas.

The rhetoric originated with a decision on terminology. In 1926, the American Petroleum Institute (API) formalized a distinction between resources and proved reserves. The former referred to all the oil known to exist while the latter denoted only stocks obtainable at a profit (Wildavsky and Tenenbaum 1981, 171–72). Somewhat counterintuitively, then, reserves lie closer at hand: they are not reserved at all. The profits they bring, of course, depend on conditions. Regulation, factor prices, or high wages might make certain ventures unviable in economic terms. Such ephemeral circumstances vastly complicated the calculation of barrels above and below what came to be known as the "commerciality threshold." Even seemingly stable conditions could shift rapidly as the industry expanded and wells reached deeper, older formations. From its inception, the binary of reserves and resources had to permit the reclassification of oil. Another term, *proving*, conveyed precisely this shift. Again, somewhat at odds with the conventional meaning, the proof referred not to the oil's presence but to its profit margin, or commerciality. An oil company might prove up resources by developing, say, drill bits capable of penetrating harder strata. Here, in fact, the industry borrowed from the language of settler colonialism. The Homestead Act of 1862 obligated newly arrived whites to show "improvements," that is, planting, cultivating, tree cutting, and so on. The government awarded titles to such investors and ejected the others, assumed to be mere land speculators. As oil supplanted agriculture in Texas, Oklahoma, and Southern California, then, new occupiers demonstrated their worthiness in established terms. In Los Angeles, wrote Upton Sinclair in the same year as the API agreement, "Dad . . . had made a big success, and *proved up* a lot of new territory, and was hailed, again as the benefactor of the Prospect Hill field" (1926, 121, emphasis added).

The preposition in this pivotal phrase opened up another field of possibility for graphic representation. Homesteaders proved "up" land in a sense that was mostly metaphorical. That elevation change hinged on modernization, civilization, and other processes imagined as linear. Once the API classification took hold, oil companies proved up petroleum before removing it. In reality, resources became reserves while in the ground. As described in language, however, the geological formation itself climbed the ladder of progress. In all these ways, the word *up* suggested a y-axis, and the API and related institutions eventually charted one. But this search for a visual vernacular encountered more obstacles than in the case of geological cross sections. Oil production depended on two variables that were easily confused: the volume of oil and its commerciality. Even if one could firmly distinguish between these indicators, institutions had to agree on assigning one to the x-axis and the other to the y-axis. After much uncertainty, Vincent McKelvey—a coworker and unstinting critic of Hubbert—provided a workable diagram (Mann 2013). In 1972, as director of the U.S. Geological Survey, McKelvey proposed a "Classification of Mineral Reserves and Resources" (1972, 35; figure 3.9). In his chart, the vertical axis corresponded to commerciality or, in McKelvey's terms, "feasibility of economic recovery." The horizontal axis measured certainty with respect to the existence of oil. This rendering clarified Proved as a category of oil both profitable to recover and known to exist. Despite these useful corrections, McKelvey's chart contained an anomaly impossible to reconcile. Shaped like a backward L, the "resources" block included a portion of economically recoverable oil deemed almost certain not to exist (in the upper right). Such conjunctures could occur. But, by including them in a corner of the graph, McElvey interrupted the sense of a linear trajectory from discovering oil to producing it. The chart suggested right turns and detours on route to riches.

Flawed as it was, McKelvey's graph stimulated better solutions in the 1980s and thereafter. In 1987, the World Petroleum Congress—which united the API with other bodies—issued a report on classification and nomenclature systems. This document presented two simpler charts—

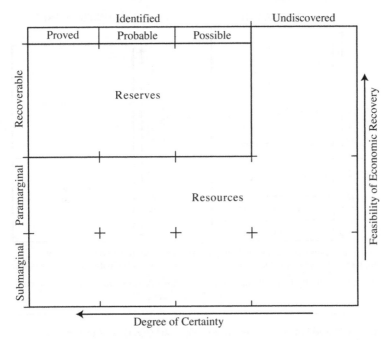

3.9 McKelvey's 1972 chart. © World Petroleum Congress. Prepared by Mike Siegel of Rutgers Cartography Lab.

almost flow charts—side by side (figure 3.10). The position of Discovered to the left of Undiscovered suggested an x-axis of certainty regarding existence. More tellingly, the left-hand column of discovered oil and gas arranged layers vertically from unrecoverable to unproved reserves to proved reserves to production. The y-axis, then, represented a commerciality threshold and even a literal flow of hydrocarbons—except that Production lay at the bottom of the chart. The vertical axis inverted geology, suggesting that oil and gas moved downward. Still, this chart portrayed petroleum in the ground and leaving the ground with enough deceptive literalism. A notation "not to scale" advised readers not to compare the sizes of the boxes (Martínez et al. 1987, 266). This warning passed into the now-definitive chart first proposed in 2001. "Guidelines for the Evaluation of Petroleum Reserves and Resources" united the recommendations of three expert bodies: the Society of Petroleum Engineers, the World Petroleum Congress, and the American Association of Petroleum Geologists. Their chart distinguishes a y-axis of financial risk—also denoted as "proj-

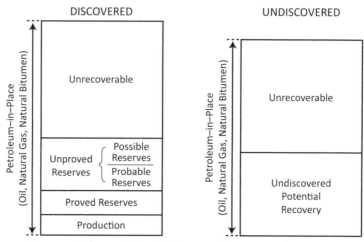

DISCOVERED   UNDISCOVERED

Reserves + Undiscovered Potential Recovery = Future Potential Recovery
Production + Reserves + Undiscovered Potential Recovery = Ultimate Potential Recovery

Illustration of basic framework of recommended nomenclature for petroleum
reserves. (Not to scale).

3.10 The World Petroleum Congress's 1987 chart (fig. 1). © World Petroleum
Congress. Prepared by Mike Siegel of Rutgers Cartography Lab.

ect maturity"—from an x-axis of existential certainty (figure 3.11). Proved
reserves lie in a corner, at the intersection of high certainty and low risk.
This chart set the standard. In 2011, the same three expert bodies wrote
it into their Petroleum Resources Management System, or PRMS.[9] Even
before that point, the chart was becoming a visual icon for the oil and gas
business worldwide. The graph constitutes what Edward Tufte calls "beautiful evidence": through it, "seeing turns into showing" (Tufte 2006, 9).

This artifice relies upon a subtle conflation of geological and commercial tropes. One might easily confuse the two y-axes: denoting, respectively, migration of geological substances and maturity of investment.
Looking very much like summarized stratigraphy, the PRMS chart super-
imposes reserves upon resources. And resources move up in the same way
petroleum rises toward the surface. A manual accompanying the chart conflates these two processes. "Budget decisions," the text concludes, "should
focus on increasing project maturity [a]s a specific accumulation moves
up the system" (Society of Petroleum Engineers 2001, 22). That sentence
shifts subtly from fiscal to lithic measures. In the latter sense, *accumulation*

RESOURCE CLASSIFICATION SYSTEM
(showing possible Project Status Categories)

		PRODUCTION			PROJECT STATUS	
TOTAL PETROLEUM-INITIALLY-IN-PLACE	DISCOVERED PETROLEUM-INITIALLY-IN-PLACE / COMMERCIAL	**RESERVES** PROVED	PROVED plus PROBABLE	PROVED plus PROBABLE plus POSSIBLE	On Production	LOWER RISK
					Under Development	
					Planned for Development	
	SUB-COMMERCIAL	**CONTINGENT RESOURCES** LOW ESTIMATE	BEST ESTIMATE	HIGH ESTIMATE	Development Pending	PROJECT MATURITY
					Development on Hold	
					Development not Viable	
		UNRECOVERABLE				
	UNDISCOVERED PETROLEUM-INITIALLY-IN-PLACE	**PROSPECTIVE RESOURCES** LOW ESTIMATE	BEST ESTIMATE	HIGH ESTIMATE	Prospect	HIGHER RISK
					Lead	
					Play	
		UNRECOVERABLE				

◄─── RANGE OF UNCERTAINTY ───►

3.11 Petroleum Resources Management System (PRMS), 2001 (the first version, published under a different title). © Society of Petroleum Engineers (SPE).

denotes a pool of oil stuck below a caprock. The hydrocarbons accumulate precisely because they are not moving up the stratigraphic system. But, in the imaginary of the PRMS, they do move from reserves to resources. Here metaphorical and literal meanings seem to fuse, as if a financial reclassification equated to changes underground. Probably few readers noticed the "not to scale" warning, presumably retained to avoid liability. Likewise, few readers reflect upon what is not in the chart: combustion and, further downstream, global climate change. In this consequence-blind, conscience-free fashion, the PRMS chart established a "zone of qualification," wherein "objects and practices are assessed to common standards and criteria" (Barry 2006, 239). A quality of upward motion reconciled the dissonant measures of commerciality and stratigraphy. Or, in Rudwick's terms, the PRMS established a "stylistic analogy," which allowed disciplines and sciences to borrow seamlessly from each other (1976, 168–69). Thenceforth, geologists and economists played the PRMS chart like a board game, forever advancing pieces toward the upper left corner.

The name of the game is "reserve replacement," and it is a sport of public relations. Reserves exit the chart on their way to refineries, fertilizer manufacturing, and so on. Meanwhile, oil and gas companies prove up resources to refill the depleted reserves. At a rate of 100 percent, replacement maintains reserves at an even level. At higher than 100 percent, reserves actually expand. Unless one discounts estimates from the Middle East, global reserves are keeping steady. In that case, the larger aggregate of resources becomes the real measure of supply, and policy makers need not worry about an impending peak. The credibility of this message depends—more than any other factor—on the commerciality threshold. Timothy Mitchell describes a "gap between the declining quantity of oil known and the quantity of expanding, yet-to-be-discovered oil" as a "space to be governed by economic calculation" (2011, 251). The PRMS chart represents commerciality as a line between contingent resources and reserves. The line is always moving. One might say, in fact, that producers push that barrier downward into the layer of resources as they prove up the affected resources. So the top stratum of resources—which is not actually higher up geologically—waits in a kind of antechamber. From 1987, the World Petroleum Conference tried to dispel this ambiguity. The authors of the report "rejected, for proved reserves, the concept of estimating future economic conditions . . . restricting these estimates to the amount recoverable under the economic conditions at the time of the estimate." Proved reserves, in other words, "are the estimated quantities, as at a specific date," irrespective of deregulation or cheaper technology anticipated in the near term (Martínez et al. 1987, 265). This effort to freeze time unraveled somewhat with the PRMS chart. That graph inserted a layer of contingent reserves just below the commerciality threshold. To make matters worse, futures contracts in the 2000s created an "oil vega," or persistent instability regarding quantities and prices (Moors 2011). Mazen Labban (2010) refers to a "financialization," where markets operate independent of oil's material disposition. Especially under these conditions, the PRMS chart allows the oil and gas business to present its raw materials as scarce to consumers but as constantly replaced to investors—without actually lying.

In Trinidad, this sleight of hand can become embarrassingly apparent as the two messages continually cross. In a small country—containing the entire commodity chain—the same firms both consume hydrocarbons and invest in their production. Having experienced the decline in oil, managing directors follow natural gas with both expertise and concern. By July 2010, proven reserves had been falling steadily, and the annual press conference—at which the gas audit of the previous year is announced—could no longer avoid this fact. Held at the Hyatt Hotel, the event began with a slide presentation from Larry McHalffey, of the Houston-based auditor Ryder Scott. McHalffey and his team had reviewed proprietary well data from every driller and condensed their numbers into the only summary that would be made public. The picture was not reassuring. A retrospective slide showed proven reserves falling from almost 20 trillion cubic feet (tcf) in 2000 to a bit more than 14 tcf in 2009. "There has been a steady decline in reserves in most categories," McHalffey related laconically. Trinidad produced roughly 1.4 tcf in 2009. Straightforward arithmetic indicated a mere decade of gas remaining. As a subsequent slide showed, the reserve-to-production ratio stood at 10. Yet the auditor chose his language carefully. He avoided linking the figure of 10 to years or any measure of time: the reserve-to-production ratio is a unitless "performance indicator," he explained. And, he continued in closing remarks, this number "goes hand in hand with reserves replacement ratio." One could not predict the future. McHalffey labeled the y-axis of his chart Project Maturity. Time was on Trinidad's side, he implied, if its leaders followed orders: "Target at least 100 percent reserves replacement each year," he concluded, perhaps patronizing his audience. The situation, in short, was imperfect but certainly remediable—as long as one overlooked climate change.[10]

The minister of energy took over exactly where the auditor left off. At that point, Carolyn Seepersad-Bachan had served for only two months—having arrived with the new government in May—and was soon to be removed in a cabinet reshuffle. She had worked in the downstream sector, as chairperson of the National Petroleum Marketing Company, and did not project confidence regarding matters further upstream. (Being female among oilmen did not help either.) At the press conference, though, Seepersad-Bachan acquitted herself admirably. Briefer than McHalffey's, her slide presentation contained no numbers or charts. Instead, she showed the map of gas acreage, stretching far out to sea. Promising, ultradeep waters billowed eastward behind Trinidad like a parachute. Replacement was coming, already begun in fact. Smiling attractively, the minister assured her audience, "We have been able to transfer exploration resources into the reserves category." And, moreover, the government was planning "continuous exploration-based activity . . . at an optimal level." The ministry would contract with British Petroleum, British Gas, and other foreign firms to drill below 5,000 feet of water. In this way, Seepersad-Barchan promised magically to "prove up supplies." David Renwick, the dean of Trinidad's energy journalists and editor of *Energy Caribbean*, spotted a contradiction. Posing a question from the press bench, he referred back to McHalffey's slide of reserves and resources since 2000. Resources in the Exploration column had fallen from 30 to 26 tcf. What did this drop bode for the future? Raised with surprising modesty, Renwick's query provoked no response from the minister or from McHalffey. Seepersad-Bachan simply repeated that she would explore more and prove up—a prediction that could never be falsified. The next day, Trinidad's highest-circulation newspaper, the *Express*, reproduced McHalffey's slide of reserves from 2000 to 2009. "Ten years left," blared the headline.[11] Had the performance failed?

Actually, no: it sustained faith in abundance among the adherents who mattered most. Despite skepticism, Renwick mostly agreed with the minister. In *Energy Caribbean*, his headline after the 2008 audit had read "Reserves Being Replaced on a Yearly Basis" (Renwick 2008b, 16). "[The] proven reserves have fallen," he admitted after the 2009 audit, "[but] the probable and possible reserves have not." That article explained the logic of PRMS in some depth: probable reserves lay above the commerciality threshold. They were just not known to exist. If they did exist, in other words, these accumulations of gas would augment Trinidad's profit-producing stockpile. Renwick converted that contingency to a certainty. "Some of that [gas] will inevitably feed into the proven category," he promised readers, "though it has not been happening as fast as industry watchers would like" (Renwick 2009, 74). Renwick and I happened to meet for the first time just the day before the 2010 press conference. Having lunch at the house I had rented, he dismissed much of the concern regarding reserves. Well versed in economics, he considered that discipline to be "the driving factor, not geology at all: the geologists just tell you where to go and drill." Hydrocarbons lay in the earth, accessible as and when funds permitted. This attitude and the audits themselves kept policy focused on when and how—not whether—to extract hydrocarbons. I suggested switching to renewable fuels. "You've got to be practical," Renwick responded gamely. "The world's economy is based on energy, and that energy is based on fossil fuels."[12] Nothing in the press conference the next day changed Renwick's outlook. To this extent then, the gathering achieved its purpose: it announced scarcity to the impressionable public while telling the experts a supply-rich story they could assimilate. Neither alternatives nor conscience ever entered the discussion. "Take [exploration] action now," the Energy Chamber urged determinedly in a letter to the *Express*, a week after the audit.[13]

By 2012, however, even loyal experts were becoming confused. In the intervening two audits, Ryder Scott had tracked a decline in gas reserves to 13 tcf. Rumors of scarcity abounded, and the industry needed to rebut them. "I know what Ryder Scott says," began Philip Farfan, "but I know nothing about Trinidad's reserves. That is quite a statement." Farfan was leading the Understanding Reserves workshop on the last day of the 2012 Energy Conference. Pacing before a slide of the PRMS chart, Farfan began by recalling the moment in 2003 when the U.S. Securities and Exchange Commission (SEC) redefined reserves. "Overnight everybody had to dress their figures down. It was a catastrophic event." Rather than proving up, they actually "wrote down" marketable assets back into the netherworld of resources. But—and here was the good news of the workshop—no oil moved physically that night. Changes to the rules proved how arbitrary—not geological—the rules were. Farfan and his cofacilitator, Tony Paul, had worked as hydrocarbon economists for decades in American, international, and Trinidadian firms. For their workshop audience of journalists and civil servants, they sought to create, rather than dispel, confusion. From the front row of that perplexed audience, Renwick asked for a comparison of the various guidelines. "We have won," Farfan responded joyously. "This is the success of today!" The bulk of the presentation tracked the fortunes of an imaginary Trinidadian gas field and gas company. Wells produced, drawing down reserves, except at three points. At those moments, the firm proved up reserves, in the first instance, by injecting water to force gas out, then by securing a new set of buyers, and last by injecting gas to bring up more gas. This story furrowed the brows of the uninitiated while tickling the genuine oilmen who had stayed around after the energy conference. "A good oil field usually gets better," interjected a jolly Texan. Farfan ended by comforting readers of the Ryder Scott report. "Don't be obsessed today with your reserves. Be obsessed with your resources." Resources, in other words, can cause reserves to grow. "Whether it is nine years left or three years left, what does that mean? . . . What matters, of course, is continuous exploration. You are bound to find something."[14]

Are you really? I asked Farfan, when we met again, nearly a year later. I wanted to know how deep his confidence ran and whether ethical scruples troubled it in any way. He had organized the 2012 workshop because Roger Packer, president of the Energy Chamber, confessed to concerns about gas reserves. "I said, Roger," Farfan recalled, "for fuck's sake get real! I'll show you why it's crap." Farfan seemed bent on personally embarrassing the SEC and Ryder Scott for imposing unfairly pessimistic expectations on Trinidad. Since 2003, proving up required a production contract. Hydrocarbons, in other words, could not be booked as underground reserves until they were at the very point of leaving the ground. What about leaving them below, I asked? Farfan parried with his professional ethic: "I feel that it is my responsibility to get as much out of the ground as possible." On the question of scarcity, he admitted to an outer boundary: "That is true about everything on the planet. I mean the sun is going to be exhausted." When I mentioned climate change, he jumped scale again: "It's hard if you're not a geologist to put the planet in perspective. . . . In a million years do you want us to be like this?" Perhaps, he concluded, we would learn from the Neanderthals, who evolved so admirably in response to global warming.[15] It was a strange—but not uncharacteristic—evasion: one that united the timescale and the inevitability of geological forces. Farfan knew his business and cared deeply about the fraternity of men and women in oil and gas. But, beyond this coterie, humanity faded into the fog of bipedal gropings toward something better. Farfan tacked from the serious to the flippant. "We got here evolving under crisis," he joked, walking toward the large car that had actually brought him to me.[16]

The Circular Back Stream

By 2010, Trinidad's oil and gas club was addressing climate change substantively—but still under an assumption of oil inevitability. The sector was beginning to treat environmentalism as an "aboveground risk," in a class with sabotage and nationalization. The risk lay in the possibility that scientists and activists might persuade consumers to cut their carbon emissions radically. In Trinidad as elsewhere, hydrocarbon insiders imagine such a scenario only with difficulty. Their term for it—"demand destruction"—suggests violent pathology.[17] As one of the milder forms of such tampering, the Intergovernmental Panel on Climate Change had considered establishing markets that would debit for releases and credit for the capture of atmospheric carbon. Having little forest with which to fix carbon in biomass, Trinidad would not benefit from substantial credits. Meanwhile, the debits would hit Trinidad's petrochemical sector and its high-emitting citizenry. To forestall this misfortune—and concurrently with the debate on gas supplies—the hydrocarbon fraternity devised an engineering project to lower the country's net emissions. They would inject carbon dioxide generated at downstream industrial sites into geological formations. This "back stream" would return carbon to its underground source, exonerating Trinidad of responsibility for those megatons (Bridge and Le Billon 2013, 37). Like all forms of what is known as geoengineering, this proposal relied upon technology barely tested (Hamilton 2013, 45–47). Risk abounded. Unanticipated losses—beginning with the Trini cow already killed by a leaky CO_2 pipe—could easily overwhelm the benefits of carbon capture and storage. Still, those advantages promised to be substantial. Through "enhanced oil recovery," the scheme would use injected carbon dioxide to push hydrocarbons up. Trinidad's back stream would reconnect to the upstream and recover unrecoverable oil. Boosters called it a win-win scenario.

Always looking for such Panglossian business propositions, the Energy Chamber began to promote carbon capture in 2008.[18] The officers of this institution saw climate change less as a moral problem than as the sort of political crisis that yields opportunities. But first they had to persuade the businessmen and politicians governing energy policy. Thackwray Driver, the chamber's British-born chief executive officer, viewed climate change with a jaundiced eye. In 1998, he had written a geography dissertation on colonials' misplaced fears of soil erosion in Lesotho (Driver 1998; cf. Driver 1999). He had also moved from London to Port of Spain, married his Trini sweetheart, and taken up work in Trinidad's Forestry Department. Among foresters, he encountered the same stale, colonial dislike of rural settlement, shifting cultivation, and large families. Bravely defending squatters, Driver (2002) challenged the long-standing imperial aversion to population growth. "I don't believe Malthus very much," he later told me, "but I hold that view very lightly: I mean I could be wrong."[19] Then he laughed. I got to know Driver—or "Dax," as he is called everywhere—quite well.

We both lived in Cascade and drove each other's kids to school. "I've always been a bit skeptical of the big claims people make," he once explained languidly on the retreating shore of Maracas Beach. "Let's muddle through and see what happens."[20] The chamber hoped that profits would happen, and, in late 2008, it drew national attention to Clyde Abder's scheme for carbon capture. An engineer, Abder mastered the details and controlled the risks. Men die on gas platforms, he informed me, in a moment's inattention. He sized his students up for potential drug use. Perhaps surprisingly—perhaps seeking a cheerier form of competence—Abder teamed up with the ubiquitous Krishna Persad.

Two years later, in 2010, Persad presented their plan to the Energy Chamber in the glittering Hyatt. This time, his humor nearly got the better of him. Persad's slide show emphasized profit, and a Midas touch. "Let me waft you," Persad invited his audience, "to a land with castles where CO_2 is transformed into black gold." Amid references to Dorothy, *The Wizard of Oz*, and streets paved with gold, Persad recruited partners to "do good and, in the process, make a lot of money." After this jarringly whimsical pitch, Persad explained Trinidad's comparative advantages in the nascent field of carbon capture: the Point Lisas Industrial Estate converted natural gas into ammonia and, as a by-product, dumped pure, high-quality carbon dioxide. Existing pipelines could carry that pollution to the oil belt, where his company would inject it into depleted, depressurized oil reservoirs. Carbon dioxide would repressurize the formation and cut the viscosity of the remaining petroleum. Like the injected water and gas in Farfan's presentation, this technology would cause hydrocarbons to flow. Indeed, oilmen in East Texas had been conducting this kind of "huff and puff" operation since the 1970s. Their wells puffed carbon dioxide down underground and huffed out oil—along with most of the CO_2. Persad would do better. By capping the well deliberately, he would trap up to half of the injected carbon dioxide for an indefinite period. As described, these back stream operations more than compensated for the resulting upstream ones: the puff outdid the huff. Oil coming to the surface, Persad told his audience, is "net cleaner than [the original] natural gas."[21] I was skeptical. Could one really capture more carbon than one emitted at the end of the day?

My search for the answer to this question began as soon as Persad stopped talking. Were carbon capture and enhanced oil recovery "contributing to the problem or to the solution regarding carbon emissions?"

I asked after his presentation. "I have no clue," Persad responded without embarrassment. In the coffee break, when the investors were no longer listening, he confessed to me, "Someone else can do the math. In that area I have no skill." On the net outcome, he revised his earlier statement: "It's neutral at best—*at best*."[22] Persad retreated even further when we met a month later in the oil belt. Inexpertly, he ran through conversion rates, admitting, "You are losing." In net terms, his back stream emitted carbon. Then he calculated some more—muttering, "My initial gut feeling is that it is negative"—and reached for the phone. "I cannot replace that carbon footprint by injecting CO_2?" he asked Abder himself. Persad listened somberly, hung up, and reported, "He said that it is not debatable."[23] I called Abder myself, and we met at his office at the University of the West Indies. "You will never have a positive carbon balance," he lectured, "by injecting carbon dioxide to produce oil."[24] The matter was obvious, in fact: gas always replaced liquid at a lower density. And long petroleum molecules unavoidably hold more carbon than does three-atom carbon dioxide. Finally, stored CO_2 could leak out later (cf. Metz, Loos, and Meyers 2005). On this issue, Persad referred me to Shiraz Rajav. Rajav had recently retired from capping wells for Petrotrin, the national oil company. Known confusingly as "well abandonment," capping could hold the carbon dioxide in place, he believed. But whether "carefully abandoned"—in the amusing phrase—or not, Persad's wells would exacerbate climate change. "[For] the amount of carbon you're going to produce in the air," Rajav reasoned, "you might as well have not put the CO_2 down there." Then, looking out from my balcony in Cascade, Rajav grieved for the Kilimanjaro glacier which he had visited twenty years earlier: "We are human beings before we are oilmen."[25] I had my answer—carbon capture was a net loser—and another question: why did most of these human, conscience-capable oilmen seem so unperturbed by that outcome?

Opportunism clouded judgment, especially at the highest levels. Concurrent with my doubtful probings, the Ministry of Energy was preparing to push Persad's scheme forward. A week after I met Rajav, as an unusually hot dry season was getting worse, the ministry announced the creation of a Carbon Reduction Strategy Task Force. The group would include Persad and Charles Percy, the current president of the Energy Chamber. Percy spoke at the press conference—held, of course, at the Hyatt—asserting that carbon capture is "the only technology known to reduce up to 90 percent of emissions from industry." The statement was true—but only if one discounted the possibility of running an industry 100 percent from solar or wind power (or not running the industry at all). Selwyn Lashley, who managed renewable energy for the ministry, might have corrected Percy. Instead, he heaped further praise upon carbon capture as "an important element in the improvement of oil production."[26] I met Lashley later at the ministry—immediately adjacent to the Hyatt—where he served as chief technical officer. Flanked by two female aides, he began by setting the task force in context: Trinidad and Tobago had "pursued this path of industrialization. It's a reality we have to deal with." He, Percy, Persad, and the rest would search for "absolute reductions in a tangible and sufficient amount." Carbon capture, I ventured, did not qualify as an absolute reduction in emissions. As I laid out Abder's quantitative, physical reasoning, Lashley and his assistants shifted uneasily. "It comes down," he responded, "to how you do your accounting and where you put your envelope." Another country would burn the oil produced from Persad's wells. "You," he affirmed—meaning Trinidad and Tobago—"produce nothing that goes into the atmosphere in your boundaries." The ministry, in short, would lower carbon emissions at the national level while raising them at the global level. "They're piggybacking on a nice concept to get business as usual done," I learned through a leak at the ministry. "It's an energy thing. It's not about the climate."[27]

Such cynicism rested on a more widely shared conceptual foundation of energy without conscience. The reigning geological-economic model made oil production appear inevitable. With or without carbon capture, the oil in South Trinidad was supposedly coming up. As Clyde Abder put it, "We gonna produce that oil one way or another. . . . As long as the world needs oil, oilmen will find it, and they will produce oil."[28] The dry season broke, temperatures cooled, and Persad invited me to a cricket match. Lugubrious as the West Indian side began to lose, he spoke about the heavy, viscous oil lurking at the bottom of his "stripper" wells. He had injected carbon dioxide into a few of them and doubled production from 4 to 8 barrels per day. He had proved up resources, I told him with a feeble smile that he well understood. But this happens anyway, he continued: "In the fullness of time, they [resources] *will* be transformed into reserves."[29] We met again socially on my follow-up visits to Trinidad but did not take up the matter of carbon capture until January 2012. By that time, the minister of energy had praised Persad's method in a speech and—in response to my question—misleadingly claimed "no net carbon dioxide will be released."[30] Persad's huff-and-puff operation was releasing 20 barrels per day. After three vodka tonics in San Fernando, I put the matter as bluntly as I ever had: why not simply keep the 3 billion barrels Persad was seeking in the ground? Ecuador was negotiating a deal to do just that with oil reservoirs under a protected rainforest (Davidov 2012; Finer, Moncel, and Jenkins 2010). Persad would have none of it. Someone else would then get the hydrocarbons, and, "You can't be held responsible if the rest of the world doesn't do what it should."[31] He reminded me of our first conversation on the topic, during which he had told me, "It's going to take a while for the world to change, and meanwhile the train is going down the track."[32] Like a commuter, the oil would arrive on schedule.

Ultimately, any policy to address climate change would have to permit the continued combustion of oil and gas. This consensus became apparent at the 2012 Energy Conference. Dax announced "Striking the Balance" as the theme. Panels on upstream and downstream concerns left little time for other issues. Climate change appeared in the interstices, almost as an afterthought. On the first day, Vincent Pereira of the Australian firm BHP Billiton raised the topic of solar and wind, in a backhanded way. "There is going to be growth in energy demand," he began blandly. Appearing open-minded, he predicted, "Renewable energy is going to grow, but," firmly now, "the reality is that fossil fuels—oil, gas, coal—are going to be 80 percent." He specified no date, but it was clear that energy production would expand dramatically, allowing hydrocarbons and renewables both to grow. What if the global movement against fossil fuels destroyed demand? I queried from midway down the Hyatt's plenary hall. Pereira responded coolly, "Having taken on challenges like this before in our industry, I have to believe that it can be solved."[33] The next day, Peter Wyant presented one of those solutions. Persad had invited him from Saskatchewan, where his agency was monitoring storage in the biggest carbon capture project in the Americas. At the province's Weyburn Field, injected carbon dioxide was delivering oil to the surface. Innocently, I popped my question about the net effect, and Wyant assured me that he was storing more CO_2 than he was releasing. Persad found me immediately after the presentation, excusing his friend with, "He misunderstood your question."[34] The next day—on my way to the reserves workshop mentioned above—I bumped into Wyant in the corridor. This time he understood my question and gaped in a fifteen-second stutter. Then he explained his assumption, really more of a wish: whoever burned Weyburn's petroleum would send the resulting CO_2 somewhere else for storage.[35] We didn't discuss whether that carbon dioxide would push out another round of hydrocarbons. In any case, the train of complicity would roll on. Seemingly, people eager to mitigate carbon emissions cannot but climb aboard.

Petroleum institutions suffer from a diagnosed "inevitability syndrome" (Nader 2004, 775). In this perverse perception, the world always needs hydrocarbons, and the substance must satisfy the demand (Huber 2012, 309; Sawyer 2010, 67). Oil and gas will themselves to come up, to be produced. Or the earth wills them to gush on top of James Dean. The text of *Giant* refers poetically to the "earth-pent oil his labors had just released" (Ferber 1952, 364). In the course of this release, petroleum experts develop an almost animistic belief in geological agency. "The oil (or gas)," writes Rick Bass, "always tries to climb higher than it is: moving, like a miner, through and between pinhead spots of porosity, trying to get up to the area of least pressure." It travels, continues the geologist from Mississippi, "back to the earth's surface, where it used to be." Hydrocarbons—like salmon swimming to spawn—fight upward to their point of origin. This notion of return, of course, emphasizes continuity more than rupture. It suggests an unbroken itinerary from vegetation to sediment to accumulation to the gas station. The voyagers pause only when impermeable layers trap migrating molecules. Bass and his coworkers drill through that rock. "Then the oil or gas . . . is just about obliged to come out. It is as daring a rescue," he congratulates his profession, "as ever there was" (Bass 1989, 27–28). The charts and categories of his profession display this operation as it takes place. They are what Andrew Barry calls "projective devices" (2013, 14). By graphing depth as time and time as depth, the geological cross section implies and predicts upward flow. Then men drill, making the prophesy self-fulfilling. The plot of petroleum always runs forward, downstream.

Why is this graphic novel so believable? It conflates geology and commerce, the natural and the artificial—and does so with particular persuasive power as regards the notion of maturation. A widely used textbook on oil defines "maturity" as "petroleum generation . . . in a source rock."

But a later chapter refers to "mature areas that have been relatively well drilled" (Hyne 1995, 172, 224). In this double use, petroleum matures as it bakes underground and as wells and pumps—and ultimately consumer demand—bring it to the surface. To describe this multiplex uplift, my informants frequently deployed the term *reality*. That conversation-stopping word imputed a monolithic quality to the petrochemical industry and its various commodity chains. The term naturalized what were merely decisions, taken every day in Port of Spain and other energy cities. The "reality" also obscured what should have been obvious limitations. Sediments hold only a finite amount of oil and gas. Demand cannot renew a nonrenewable resource. Yet experts assume that hydrocarbons will not run out as long as governments, corporations, and people don't want them to. Tyler Priest, in a history of Shell Oil, captures this fuzzy thinking unwittingly. He writes of the exploration team going offshore "in the race against depletion" (Priest 2007, 106). In the real reality, they were racing toward depletion. In the reality of charts, however, Shell proved up resources so fast that reserves grew. There is no Malthusian absolute scarcity in this economic thinking. Meanwhile—and of more immediate concern—the atmosphere is imposing its own limitations. The oil fraternity has trouble focusing on this fact. "The Earth has become abstract," writes Bill McKibben, "and the economy concrete to us" (1989, xxiii).

Individuals may have also faded into a strange obscurity—even within the macho hydrocarbon industry itself. "Oil is found in the minds of men," my informants occasionally told me (cf. Yergin 2011, 717). But when the conversation turned to climate change, even the most accomplished geologists felt oddly powerless. In Trinidad, most understood the way in which hydrocarbons were changing the climate and threatening life. Many wished they could do something about it. It took me some time to comprehend this form of passivity and complicity. For social scientists, climate change has thrown human agency into stark relief. High emitters of carbon dioxide now stride across the planet as "geological agents" (Chakrabarty 2009, 206). Yet the experts of oil have not yet adjusted to this inversion of deep and shallow time. Krishna Persad cared enough about climate change to create a solar-powered eco-resort on Tobago. Not merely an object of greenwashing or corporate social responsibility, the family-run enterprise occupied his thoughts and moved his spirit. Why, then, did he at the very same time explore for new supplies of oil in South Trinidad? If not him, someone else would, he always said, as if submitting to an unwelcome fate. Finally, I put it to him that he was the best geologist in the country, a mind that regularly found oil. He also took greater economic risks in enhanced oil recovery than any other producer (Renwick 2008a, 20). Just possibly, oil he did not personally discover or flush out might stay there, and that oil trapped in sediments might keep the world within the safe climate boundary. Persad wavered, taking the scenario seriously: "If I was going to be responsible for tipping the whole world over, then I can't do it."[36] He brushed against the boundary of conscience.

CHAPTER 4

Lakeside, or the Petro-pastoral Sensibility

I saw Trinidad's Pitch Lake first in the National Museum (figure 4.1). An 1857 watercolor hangs on the second floor, in a kind of shrine to the island's first professional painter, Michel Jean Cazabon. Cazabon, the son of wealthy French planters, helped launch a national consciousness among the island's polyglot, multiracial population. Through the brush, he converted a space hardly known or valued by its own inhabitants into known, named, beautiful places (Cudjoe 2003, 154). Thus, the pleasant watercolor titled *Asphalt Lake* stands as an exception to my general finding of symbolic thinness regarding hydrocarbons. The image is still more improbable. As I discovered upon visiting the Pitch Lake, Cazabon saw a scene both banal and unpretty. At 114 acres (48 hectares), the feature resembled a large, puddled parking lot. I toured the lake dutifully and then looked for people in La Brea, the adjoining town named in Spanish after pitch. Activists were fighting a proposed aluminum smelter, a threat to the environment and human health. I had met these critics in Port of Spain and then renewed my acquaintance at their homes and in political meetings. Were these working-class Afro- and Indo-Trinidadians concerned about the current danger from oil as well as about potential toxins from a smelter? I probed in this way and hit a wall. To a person, these activists—as well as their allies in Port of Spain—expressed a positive view of hydrocarbons. Without citing the painter, they shared Cazabon's rosy of view of asphalt, petroleum, and gas. I continued my quest. On repeated visits of a day or two from the capital, I crisscrossed South Trinidad looking for dissenters. Trinis were preparing for Carnival, but I hoped to find, at least, someone concerned about global and local oil spills. Yet as I moved east and north from La Brea—passing Cazabon's home region of Naparima—I encountered complicity and silence. In the oil belt, I met many earnest, environmentally minded people but no kindred spirit.

4.1 *Asphalt Lake*, watercolor, Michel Jean Cazabon, 1857. © Michel Jean Cazabon, Courtesy of the National Museum and Art Gallery.

Oil's natural quality shields it from a degree of interrogation and fear. Remember, petroleum geologists believe that 95 percent of all the hydrocarbons ever formed in Earth's crust have come up without human intervention (chapter 3). People extract the small, stubborn remainder and—crucially with respect to the climate—burn all those stockpiles. Combustion releases carbon dioxide, most of which, again, already exists in nature. The entire hydrocarbon fuel system, in other words, generates no new substances. Plastic manufacturers do convert oil and gas into artificial compounds. But on the energy side of downstream production, refineries and power plants generate nothing so exotic as the radioactive isotopes minted in any nuclear station. Human actions, therefore, only amplify petrolic activity that preceded industrial life. Of course, that amplification bursts the bounds of moderation. Global climate change makes that clear. Locally, in the Niger delta, jagged-edged pipelines and lurid gas flares provoke rebels, writers, and photographers alike. With a similar starkness, the vast strip mine amid Canada's tar sands overruns boreal forest. Observers near and far criticize these oil fields as environmental and social atrocities. Less extreme cases attract less criticism than one might expect—nearly none, in other words. In Texas, Oklahoma, and California, pump jacks sway like horses, a metal rodeo bucking up oil. After so many decades, they seem to belong amid tumbleweeds and alkali soil. The literary scholar Leo Marx names this aesthetic compromise the "technological sublime" (1964, 230). Americans have applied it in appreciating railroads, bridges, and sky-scrapers (Nye 1994). Oil wells too: in the eye of the tolerant beholder, rigs thrum gently, undisturbingly. Offshore, fish and aquatic vegetation colonize production platforms, as the Houston aquarium is eager to display (Jørgensen 2014, 267–68). To the resident, this oil land is home: a mosaic of water, soil, plants, animals, and a gooey mineral widely sought and contentedly consumed (Campbell 2014, 83, 101).

Such an imaginary—which I call the petro-pastoral—shaped my entire experience in South Trinidad. In this chapter, I relate the "infinitely malleable" pastoral genre to cultural sensibilities among Trinidadians who actually write very little (Garrard 2004, 33). Outside Trinidad, pastoral authors have long applauded small-scale, communitarian alternatives to the capitalist, technological city. In villages, one should live a less regulated life, keep one's own hours, and marry for love. The English poet William Wordsworth wrote famously of the beauty of Cumbrian landscapes filled with lakes and interspersed with fields, forests, and meadows. These aesthetics run like a red thread through contemporary England's rural nostalgia and frequent romance with agrarian life (Williams 1973). Trinidad is different. Even without oil, this sort of conventional pastoralism would hardly enter into its literature or calypso (Rohlehr 1992, 203). Caribbean farming has frequently evoked brutal, rather than gentle, times (Deloughrey 2004, 299). Sugarcane, the island's chief crop, carried the taint of slavery and indentureship long after the abolition of both institutions. Amid that legacy, the country's first president, Eric Williams, industrialized with a palpable "scorn for agriculture" (Miller 2011, 54). He applied nostalgia—or, at any rate, assumptions of rightness and belonging—to the energy sector. Inflected by the "petro," Trinidad's pastoralism applauds the oil well, rather than the village water well. Even when the well damages ecosystems and human health, petro-pastoralism underwrites a surprising tolerance. That indulgence frankly frustrated me—but not enough to impede my ethnography. On the contrary, I kept moving through South Trinidad, trying to find an antioil sentiment. Stubbornly, I sought a conscience about hydrocarbons and climate change in the most unlikely place. Meanwhile, movements complacent toward hydrocarbons brought the party founded by Eric Williams to its knees. One government left office, and the other replaced it, but no policies or practices changed with respect to hydrocarbons. I lived through a historical turning point that—because of the lakeside aesthetics of La Brea—failed to turn.

Defining the Lake

Hydrocarbons straddle the boundary between two forms of nature: the biosphere and the lithosphere, the landscape and its rocky substrate. From antiquity, naturalists have favored the surface. Among the first geographers, Hippocrates titled his most influential work on the environment *Airs, Waters, Places* while Strabo later focused inquiry on inhabited space (Glacken 1967, 80–81, 103). The underworld remained deeply foreign. That sense of strangeness—surviving religious and philosophical shifts far too complex to describe here—allowed most Europeans simply to forget the lithosphere. Even in the modern period, painters of nature depicted rocky mountains rather than deep rock (Schama 1995, 385ff.). In fiction, Jules Verne did send Professor Lidenbrock on his *Journey to the Center of the Earth* (1864). Scarcely imagined since, that subterranean trek still seems original. This sort of surface bias has long run through popular and academic discourse in North Atlantic societies and their colonial settlements (Scott 2008, 1857). The subsurface, after all, lies behind a veil, and literal-minded folk draw, describe, and remember what they see. Oil, however, breaks into view; it is one of the few aspects of deep geology to do so regularly and in many parts of the world. Seeping slowly through a fissure or spurting out through a well, hydrocarbons demand an explanation. Are they vegetable or mineral, topographical or geological, a product of soil or of deep rock? The Pitch Lake—which appears in Charles Lyell's (1830) foundational *Principles of Geology*—received perhaps the earliest and most thorough scrutiny of any single petroleum deposit. Ultimately, that debate would determine whether La Brea stood in the vertical column or on a horizontal plane. Did it perch perilously atop an unfathomably deep volcano or comfortably alongside a pastoral lake?

The extended first round of this enquiry reached a reassuring topographical consensus—but not before considering other, more outlandish possibilities. During his 1732 visit to Trinidad, the Jesuit Joseph Gumilla encountered pitch as well as the indigenous people living near it. The substance appeared to him like a "spring" that ran "inexhaustibly." The neighbors appeared to live in terror. Just before Gumilla's arrival, "a piece of land had sank . . . and then in its place had appeared another pond of pitch, to the fright and fear of the residents, suspicious that, when they least expected it, the same would follow inside their settlements."[1]

A century later, Lyell concurs with this Amerindian account. In his view, "the frequent occurrence of earthquakes," "volcanic action," and "subterranean fire" both produced petroleum and ejected it to the surface (Lyell 1830, I:218). Not long after this observation, George Wall and Jas Sawkins (see chapter 3) disputed it. The two men bored into the pitch lake— through the Pitch Lake, in fact—to a bed of clay. Sawkins drew the deposit as a set of convection cells resting on a "supposed solid surface" (figure 4.2; Wall and Sawkins 1860, 140). Wall's main text concluded unambiguously, "The origin of the asphalt, is in the stratum itself, and not referable to any process of distillation or ascension from below" (Wall and Sawkins 1860, 143). Asphalt did not flow into the lake, then; it came into being as the lake converted the remains of plants and animals into more asphalt. Six years later, Wall described the "direct production of bitumen from vegetable remains" (1866, 236). Tropical air at 80°F simply baked rotting vegetation. There was no underground mystery, he concluded: "The generation of bitumens is easily explained by the operation, at ordinary *terrestrial* temperatures, of chemical laws" (1866, 239, emphasis added). Like most aquatic lakes, pitch simply occupied a depression in the ground.

That comforting judgment held—until fears of a pitch shortage reopened the entire issue. Conrad Stollmeyer and others extracted and sold asphalt so rapidly that the lake began to shrink (see chapter 2). In 1901, the colonial government convened an Asphalt Industry Commission, but before it could issue its report an upstart geologist published his own short pamphlet. Actually trained as a civil engineer, Oscar Messerly disputed the reigning description of the lake as "simply a large puddle of pitch which has oozed out of the sandstone and collected in a basin-like depression in that rock" (1902, 18). Had Wall and Sawkins drilled in the right place, they surely would have found deep "chimneys." These fissures, which Messerly sketched (figure 4.3), linked the Pitch Lake to "an enormous quantity of organic matters" deposited beneath the Gulf of Paria as the Orinoco scooped that "meditaranean sea [*sic*]" into existence (Messerly 1902, 8). Bitumen

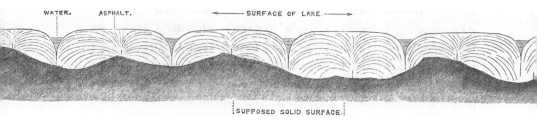

4.2 Wall and Sawkins's cross section of the Pitch Lake, 1860.

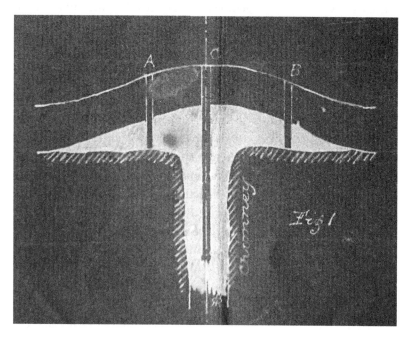

4.3 Oscar Messerly's chimney in the Pitch Lake, 1902.

then rose through the chimneys to "orifices," which included the Pitch Lake and similar features in Venezuela (30). Extraction appeared to prove this point. As one dug pitch, Messerly observed, the substance flowed in to fill the hole. It toppled in from the sides, certainly, but also welled up from the bottom of any hole or cavity. "Vertical pushing" smoothed the asphalt in La Brea (19). Were it not so, depressions would remain at the sites of extraction. The earth's crust, in other words, always replenished what one digger had taken from another. This upwelling suggested new nomenclature. Messerly chided those authorities who had "very improperly given the name of Lake, as it offers not one of the topographical conditions which may justify such a denomination" (5). More likely, powerful subterranean forces operated below La Brea. In his brash, undiplomatic way, Messerly almost made the catastrophic theories of Gumilla and Lyell fashionable again.

Then the Asphalt Industry Commission buried Messerly and his geological theory in a thousand pages of transcripts. Doubt reigned throughout the hearings. Consisting of a geologist and two barristers—and attended by the attorney general and six other legal men—the commission first interviewed a series of asphalt professionals. It spent the better part of a day, for instance, with Mr. G. F. Bushe, an agent responsible for Stollmeyer's portion of the lake. Bushe initially echoed Messerly's view of vertical pushing. In any given hole, he testified, "The amount that comes up in three months' time is too large to come from the sides, and therefore it comes from below" (Colony of Trinidad 1903, 45). The next witness discounted all forces but subterranean ones: "The whole pressure comes up from the bottom," claimed Arthur Protheroe, the owner of a lake lot. "You can see it rising." The attorney general led the witness: "You believe there is some underground supply?" "I believe so," Protheroe rejoined (65). He had watched diggers excavate to an apparent clay basement. Then the clay rose up, more than a foot in a night. Protheroe had supervised another 20-foot hole. "Did you get to the bottom of the pitch then?" pressed the attorney general, Nathaniel Nathan. "No, Sir," answered the witness (74). Interviewed on the next day, Francis Duncan claimed to have dug down twice as far. The pickaxes cut through clay and encountered more pitch. "Did you get to the bottom of that pitch ever?" asked Nathan (127). At 40 feet, Duncan had still not exited the stratum of bitumen. Finally—for this phase of the hearings—the pitch digger Alfred Rogers conjectured about a structure akin to Messerly's chimneys: Asphalt "comes from vents, as it were . . . all the time" (137). Clearly unconvinced, Nathan derided the witness's "constitutional impossibility of measuring lengths of time" (131). Some commissioners had already made up their minds.

That skepticism toward the diggers eventually won out. After twenty-two grueling days of testimony, Nathan and the others concluded that pitch was shallow, a topographical form. The commissioners explicitly endorsed Wall and derided Messerly. Although they rejected Wall's model of ongoing asphalt production from vegetation, they characterized pitch emphatically as superficial (Colony of Trinidad 1902, 7). No "earth pressure" pushed pitch through chimneys, asserted the garrulous attorney general late in the hearings. "I think we may all assume," he declared rather heavy-handedly, "that pitch was formed *in situ*, where it was found, and is not a volcanic product" (Colony of Trinidad 1903, 345). With evident scorn, the commission's final report referred to Messerly's ideas as "too visionary to need examination" (Colony of Trinidad 1902, 12). Had all the witnesses who concurred with him suffered from similar delusions? No, the commissioners explained, they were merely missing the forest for the trees. Shallow hydrostatic pressure could push up in a fashion indistinguishable from deeper geological pressure. Asphalt surrounding a hole behaves rather like water pressure in and upon a cup pushed underwater. Or, as an amenable witness had illustrated, "Here's my hat. It has a flat bottom. Press the sides, and crown goes in" (Colony of Trinidad 1903, 203). This liquid property, then, gave the impression of upwelling and increase, when pitch was finite and actually decreasing. On this point, the commission ruled definitively: "The amount of asphalt in the lake and the land deposits is strictly limited and . . . the surface level of the deposits must be lowered by any mining upon them" (Colony of Trinidad 1902, 15). Pitch was falling rather than rising, a lake receding from its shore.

Largely forgotten now—even in La Brea—the Asphalt Industry Commission made history by what it did not say. It did not describe pitch as a deposit, outcrop, seep, or any other geological term in contemporary usage. It did not label bitumen as a pollutant from the underworld. Speculate for a moment on the counterfactual: the commission could have traced bitumen to Verne-like deep structures and processes. Perhaps Trinidadians would have read and reported this finding as an affirmation of the Amerindian myth. Perhaps La Brea residents would have treated the Pitch Lake and its products with greater caution. Like Messerly, they might have interrogated its designation as a still-water lake at all. A more scientific version of the story was soon proved true: pitch did seep up from deep below. Over the course of the first half of the twentieth century, petroleum geologists reached consensus on the migration of hydrocarbons from deep in the crust (Kropotkin 1997). Messerly had got it right after all. But this reversal in the dominant judgment came too late to disturb the hard-won consensus. In the wake of the Asphalt Industry Commission, practice established the Pitch Lake's reputation as a purely terrestrial, topographical body. Tourists visited, walking carefully so as not to sink into the liquid, sticky parts. Like many, I bathed in the healing, sulfurous waters lying in its crevasses. As popularly understood, the landscape cupped and contained its dark pond. Almost in Wordsworth's pastoral sense, La Brea became a lakeside village.

The Nonindustrial Industry

If, as Raymond Williams writes, rural nostalgia contrasts capital with community, La Brea created useful history of both kinds in the twentieth century. In the wake of the Darwent well (see chapter 3), oil production spread throughout the southern tier of the island and to the Pitch Lake. La Brea became an oil town—as well as a pitch town—surrounded by the infrastructure and staff housing of British and American oil companies (Higgins 1996, 180ff.). Wells, oil tanks, and ponds of water filtered from oil dotted the environs. The industry cut many corners in those early days. The infamous Dome Fire of 1928 killed sixteen workers and bystanders (de Verteuil 1996). Outside the actual fields, all the companies practiced blatant discrimination. They limited the rise of black and Indian workers and squeezed them into crowded, segregated accommodations known as barracks. Frustration eventually boiled over in strikes, riots, and sabotage in 1937. By midcentury, however, the industry seemed to have overcome these growing pains. It no longer caused visible, violent damage. Trinidadian workers advanced, gaining in expertise and managerial positions. Meanwhile, as its oil fields were depleted, La Brea suffered a slow decline. Natural gas, liquefied nearby in Point Fortin, became the country's leading natural resource. Still, La Brea did not go bust. Unlike its codiscoverer of petroleum—Titusville, Pennsylvania—South Trinidad stayed in the hydrocarbon business (Black 2000, 189). The town continued to bring up pitch and some oil, leavening the technological sublime with a measure of nostalgia. Born in 1928, Arthur Forde recalled the La Brea of his youth as the "industrial capital of the Caribbean": "We were more modernize [sic] than any other village in Trinidad."[2] Misty-eyed, his memory called to mind a hamlet of machinery—and reconciled all the apparent contradictions in that sensibility.

At his snack shop—across from seeps that sometimes burned spontaneously—Forde also gave me a lead. Agatha Proud, he confided, once owned the Pitch Lake. Chasing these rumors became my obsession. People referred me to Ethelbert Monroe, another elder inhabiting one of the many vintage houses tilted wildly in pitch-infused soil. Monroe remembered "Miss Proud" as a "Negro woman" who owned "the whole of the Pitch Lake." She also claimed the adjacent parcels, belonging legally to oil companies and rented to tenants. She visited the latter, demanding rent and "threatening to move them from the land."[3] Errol Jones, who sculpted bitumen into tourist art, corroborated these details and added more: Miss Proud had worked as a servant for an American family, who pretended to possess property claims to the Pitch Lake. Dying without offspring, the family willed its supposed entitlement to Miss Proud. She pursued that claim until a fire—perhaps due to arson—destroyed all her documents. I believed this version, but Jones concluded, "They say she was screw-loose."[4] Other residents recalled Miss Proud as more sane and more deeply rooted in La Brea. Proud was "probably one of the last of the natives living here," recalled a guide at the Pitch Lake.[5] The woman, who was likely born around 1900, died or disappeared in the 1960s. No further facts were available. Although she had pressed her claim in court, I could find no record of it. Finally, I tracked down a neighbor of Miss Proud, born in roughly 1920. (She did not know the exact year.) Over tea and biscuits, Virginia Piper spoke as vociferously as her frailty allowed: "This t'iefing company . . . t'ief Miss Proud." Piper was referring to Trinidad Lake Asphalt, the current holder of Stollmeyer's original concession. Proud, she continued, had inherited the land from her grandparents, at least some of whom were Amerindian.[6] Piper grew agitated recalling injustices recent and antique. Perhaps, Proud's ancestors had met Gumilla.

As I found, Piper's account crossed the generations in La Brea. On another visit, I walked along the main road, also buckling from pitch migration, to the house of Joshua Logan. Handsome and in his twenties, Logan taught drama at Vessigny Secondary School. While earning a bachelor's degree at the University of the West Indies, he had written a one-act play about the Pitch Lake called *The Price of Progress* (Logan n.d.). The story centers on the Bird family: a man known as Bird and his deceased great-great-grandmother. This ancestor, called "Mamma" by all her descendants, represents Agatha Proud. A French planter owned Mamma. When his child falls desperately ill, he asks the slave for an African remedy. Upon the cure, as Bird recounts in Scene II, the master frees the slave and grants her a wish: "Mamma say she want de Pitch Lake," narrates Bird, and she gets it. Why—so long before Stollmeyer and others began selling it—did Mrs. Bird want the asphalt? "It had healing powers," the younger Bird continues, and Mamma "wanted de Pitch Lake for everybody." Later, she refuses to sell to extractors because "people does use it, and it is sacred." When we met at his house—just across the street from Monroe's—Logan explained the avian reference in Mrs. Bird's name. He wanted to link Miss Proud with the Amerindian myth of the Pitch Lake and with "that oneness with nature."[7] The play thus represents hydrocarbons as nature's gift to wellness—stolen by capital. Avaricious firms collude with the state to swindle the Bird family. Just as, in Logan's experience, oil companies pollute the water and destroy the local fishery. But the penultimate scene undercuts the author's own criticism. Bird dies suddenly and bequeaths a small fortune to the fisherman whose livelihood is most threatened by pollution. It is a win-win: the worker need no longer work, and capital can continue business as usual. Tempered in this way, Logan's drama advanced to the national Best Villages competition.

Meanwhile, the real-life toxic drama of La Brea appeared to be heading toward a less happy ending. Logan's uncle, Noah Premdas, led an environmentally minded group called La Brea Concerned Citizens United.[8] Implacably, the group opposed the government's plans to build an aluminum smelter. Meanwhile, these activists took no stand against hydrocarbons. I first encountered their paradoxical sensibility one evening in 2009 after a meeting of the group. Liming (hanging out) in Premdas's open garage, the activists recalled the violent destruction of their forests and lakes. In 2004, before the government even imagined siting a smelter in La Brea, the state had ordered bulldozers to flatten the trees and workers to club the fleeing animals to death (deGannes 2013, 31; Sheller 2014, 218). Up to that point, people had recreated in the woodland, often, as one activist put it, "giving thanks for the virgin environment, the untouchedness." This virginity still accommodated quite a bit of touching. When municipal water supplies failed, as they frequently did, families had collected their drinking supplies from nearby ponds. They drew a ring on the water with soap, recounted his colleague, "the soap used to push the oil [away from the water being collected]. . . . They were like little scientists. They were experimenting."[9] At that, everyone giggled. The site's three lakes, I later learned, resulted from the dumping of "produced water," the highly contaminated fluids separated from oil. Locals knew this history of the 1930s: they called the pond with the most obvious sheen the "oil dam." An environmental impact assessment—carried out before the bulldozing—showed levels of oil and grease 8.7 times Trinidad's health standard.[10] In effect, my informants treated hazardous waste as a minor irritant, like sand blowing on their beach picnic.

If oil in water caused no alarm, the infrastructure for producing oil almost provoked celebration. In La Brea, the site cleared for the smelter had contained derelict oil storage tanks and twenty-seven petroleum wells. When the government capped those wells in preparation for construction, onlookers misinterpreted the operations as more drilling. They greeted this development with equanimity. "It is only when people started to see the magnitude of the clearing," Premdas recalled, that they realized a more sinister project was afoot. Premdas, in fact, worked as a well survey supervisor for Petrotrin, the national oil company. "People actually live with— you can say—oil fields in their yards," he explained, "a few feet from their houses." He defended oil and attacked the smelter by saying, "We for any industry that doesn't create a health risk to the communities."[11] Of course, he was also for an industry that employed his neighbors and himself. But the other activists—as I spoke with them before and after the meeting— nearly overlooked the economic benefits of oil. They remembered life with oil, rather than a living from oil. A Rastafarian, Isaac Gregory wore his hair in dreads and considered the earth sacred (figure 4.4). He put me up for the night after the meeting, gave me breakfast the next day, and submitted to my questions on the pump jacks that used to suck oil from nearby wells. "There was no hamburg," he responded, using a Creole word for *problem*. "We used to get up and ride on those," he continued with amusement. "They used to look like horses."[12] Adam Chalant smiled at the same diversions: "I always like[d] the area . . . nice, quiet, serene. You could do what you want." Did the whirring of pump jacks disrupt that quiet? I asked. No, he replied, "You'd see more or less a puddle of oil bubbling. It wasn't, to say, dangerous."[13] We sat at the protesters' encampment, just outside the gate of the construction site. Their signs labeled aluminum a "death industry," and one—perhaps written by Gregory—referred to "Smelter Babylon." They also labeled it a "heavy industry" and vowed to keep it from a community and ecology otherwise at peace.

To me, however, bitumen seemed the heaviest of all the products in question. How could one assimilate the mining of pitch to the activists' bucolic image of their locale? After some difficulty, I gained access to Trinidad

4.4 Isaac Gregory, 2009. Photograph by the author.

Lake Asphalt. Having exported material for major bridge and tunnel projects worldwide, the company had recently perfected the bitumen pellet. In this bullish atmosphere, the firm hosted its annual calypso competition just before Carnival. I bought my ticket and, as a foreign curiosity, soon found myself in the deafeningly loud VIP section. While amply wined and dined, I heard two songs mentioning the smelter. Alfred Antoine's lyrics sympathized with a woman protesting aluminum, regretting that "industrialization, it come to stay."[14] I followed up with him. A casual worker— possibly uneasy in the head office where we later met—Antoine backed away from any criticism of the smelter. "I for industrialization," he assured me, but he associated the term with future projects only.[15] Trinidad Lake Asphalt fell outside this category. The second calypsonian, Roger Achong, worked as a well-paid chemist. His calypso praised "Mother earth [who] will bless and see you through" and pronounced the smelter "an environmental blight." At the same time, he endorsed "fires of progress . . . burning bright" and criticized oil and gas only for their depletion.[16] In our conversation, Achong clarified the distinction between pitch and aluminum. "It's a natural project," he assured me, gesturing toward bitumen samples in his office. "We are like caretakers." Achong contrasted this stewardship with his earlier experience working in Trinidad's methanol plant, the largest in the world. That industry synthesizes ammonia gas at intense temperatures and pressure (Hager 2008). There, "you have a chemical change taking place from the beginning to the end," he explained. "This [Trinidad Lake Asphalt] is just a mining operation with a still. It's a joke."[17] Laughably simple, the still heated asphalt to 150°C—boiling away the water—while maintaining atmospheric pressure. Asphalt, in other words, was less industrial than moonshine and equally compatible with rural life.

As critics of the smelter reiterated their tolerance for oil, I scouted farther for someone willing to criticize both of them. Where was the conscience of energy in Trinidad? Searching for it, I followed the hydrocarbons from upstream production to sites of midstream refining and downstream manufacturing. Just east of La Brea lay San Fernando, the unofficial capital of the oil industry and a town large enough to produce its own Carnival parade. Most revelers bought masquerade costumes from a small number of producers, chief among them the award-winning Kalicharan family. In early 2010, the Kalicharans introduced eight costumes. Focused on the theme "Outta d'Rain Forest," the skimpy outfits evoked macaws, leopards, and similarly colorful creatures. This biodiversity harbored one oddball, however: a blue-black, fringed bikini advertised as "oil spill." Thinking I had at last found an antipetroleum artist, I took the ferry from Port of Spain to San Fernando. No, Wendy Kalicharan explained, the slick-like bathing suit did not indict hydrocarbons—sponsored, as it was, by Lenny Sumadh, Ltd., Automotive, Petroleum, and Industrial Supplies. Mrs. Kalicharan only objected to oil spurting, or "sputing," as she put it. Producers "need to be a little more environment friendly and look into that," she said, before referring me to her daughter, the real mastermind of the costumes.[18] Ayana Kalicharan worked for a hydrocarbon firm on matters of health, safety, and environment. Her employer, she assured me, was "going green," as were many others.[19] The ambiguous costume, she clarified, referred to oil spills elsewhere, not in Trinidad's rain forests. Ayana had studied Occidental Petroleum's horrific spills in Ecuador and used these events as a cautionary tale in the brochure distributed at Carnival (cf. Sawyer 2004). "Fossil fuels," promised the Kalicharans in that document, "can be extracted in an environmentally friendly way."[20] Once again, complicit parties gave blue-black crude a whitewash.

Still seeking a nature-minded criticism of oil, I continued slightly north of San Fernando. Since Molly Gaskin founded it in 1966, the Pointe-a-Pierre Wildfowl Trust had operated on the grounds of the country's oil refinery. In that facility, pipes snaked everywhere, as gas ignited in flares. To me, the lurking smell of sulfur—and constant warnings against fire, sabotage, and even taking pictures—blared danger at every corner. I drove through Pointe-a-Pierre vigilantly. Gaskin felt these sensations too. But, as she explained to me at one end of the bird pond, "once you are inside here, you don't know that there is a refinery out there." The abundant avifauna did not appear concerned either. Their water, Gaskin continued, had not spurted from an oil well. Rather, the refinery had pumped up liquid from aquifers and stored it in ponds to cool the machinery. Water waited, so to speak, upstream before—rather than downstream after—an encounter with hydrocarbons. Still the entanglement appeared to shape Gaskin's efforts at environmental education. I was struck by a poster on the pond's shore that mentioned "industrial pollution in the atmosphere, including gas and oil." Was a conscience-stricken Gaskin blaming Petrotrin for climate change? No: she disclaimed the billboard as a donation from the Organization of American States. She would have modified the indictment of oil and gas with a caveat regarding "proper checks and balances." On her own account, Gaskin had printed a poster, "What You Can Do," which recommended mildly that one inflate one's car tires fully and turn off unused electronic devices. A "balanced" smelter might proceed too because, as Gaskin elaborated, "if you are totally unreasonable . . . you don't make any sense."[21] Among industries, only the nuclear sector lay truly beyond the pale. Gaskin had first made news in 1995 when she invited Greenpeace to help protest the transshipment of nuclear waste through Trinidadian waters. Her elastic environmental sensibility embraced any pollutant short of radionuclides.

A final push northward took me outside the oil belt, to the Point Lisas Industrial Estate—and to the most full-throated endorsement of hydrocarbons. Amid factories heaving petrochemicals, I did not expect to find critics of industry. Engineers employed at Point Lisas saw oil—and aluminum too—as beneficial. Indeed, hydrocarbons, they believed, might neutralize the most severe threats to health and the environment. I found

Reeza Mohammed in the office of the National Energy Corporation. For that parastatal firm, he was coordinating environmental projects, including some proposals in the vicinity of the smelter. With a doctorate in physio-pathology, he had served in 1999 and 2000 as Trinidad's first minister of the environment. I asked Mohammed about water quality in La Brea. Given the high levels of oil and grease in the reservoirs, why had residents not succumbed to widespread illness? Mohammed, who warmed to his topic of expertise, misunderstood my question. He was concerned about fecal coliform bacteria: the water "was so bad, if you put a spoon in it, the spoon was standing on its own." "For the life of me," he continued, "I could not understand" how people avoided diarrhea. Oil and grease might have actually restored gastrointestinal function. "The seepage of hydrocarbons in the area," Mohammed speculated, "would have affected [reduced] the pathogenicity of the bacteria."[22] Other experts ridiculed his theory, but the model of oil as salve applied more widely. Douglas de Freitas ran a company specializing in bioremediation. From offices close to Point Lisas, he oversaw the cleanup of contaminated sites in La Brea and elsewhere. "It's the perfect site," he enthused about La Brea. At 20 or 30 feet below ground, pitch provided "a natural sealant." "We have those massive tar sand deposits below us," he boasted to me. As he advertised to potential clients, nothing would leach through the town's bituminous shield.[23] Hydrocarbons protected people and the environment from contaminants far worse.

At root, industries generated dread in and around La Brea—and offended conscience—only when they involved fast, high-risk processes. The industrial sociologist Charles Perrow distinguishes such operations from mere mixing, separation, or fabrication. Thoroughly artificial, these transformations occur at high temperatures and pressures, accelerating beyond humans' ability to monitor and control. Fission, as Perrow's (1999, 9–10) quintessential case, sustains itself independently in a nuclear power station. Aluminum stands closer to such self-piloting than to tool-like behavior. In a smelter, raw alumina flows into room-sized pots where it is bathed in cryolite and aluminum fluoride, heated to 1000°C, and submitted to powerful electric currents. This Hall-Héroult process creates aluminum while coating the pots with a highly hazardous residue and sometimes releasing dangerous vapors of hydrogen fluoride. In the event of an accident, human operators can stop the reactions, but not immediately. Some of the well-educated activists from Port of Spain understood these specifics. More vaguely, members of La Brea Concerned Citizens United feared smelting as a kind of juggernaut. Adam Chalant defined heavy industry as making noise "all day all night" with "shifts." In light industry, "you have an opening time and a close off time." Trinidad Lake Asphalt, whose equipment slept at dusk, harmonized with nature and the land in this way. Down the peninsula, residents of Pont Fortin drew even finer distinctions: some waxed nostalgic for the blue-yellow flare of the refinery, replaced by the orange flare of natural gas liquification (Campbell 2014, 63). In La Brea, from where one can glimpse all these flames, an activist still contrasted this local "scenery" with the unwelcome "bright life" of Port of Spain.[24] In the bucolic night, Chalant had even seen the deity known as Papa Bois.[25] Having survived since the French planters, this forest spirit now presumably flits among the pump jacks of South Trinidad. Ultimately, the specter of aluminum made oil appear even safer than people already considered it to be. For residents of La Brea and the oil belt, petrolic pastoralism involved no contradiction in terms.

The Island against the Mega

At the national level, the movement against smelters began with tropes closer to standard, agrarian pastoralism. In 2006, the government first suggested manufacturing aluminum at Chatham (deGannes 2013, 3). Down the Cedros Peninsula and past La Brea, this village lay beyond the arc of industrial sites. Indo-Trinidadians raised crops and fished in a string of inland settlements and beachside villages. They occupied a patchwork woodland described in Jamaica as "ruinate" (Cliff 1987, 1; Maisier 2015, 117). The present seemed to constantly churn through the past. This temporal topography fit the popular "racial landscape" of South Trinidad—where descendants of indentured laborers still worked the land (Khan 1997, 4). Prime Minister Patrick Manning and his mostly Afro-Trinidadian cabinet seemed bent on industrializing that village life. Then a fortuitous discovery both deepened the heritage of Chatham and gave it African roots. In August 2006, archaeologists uncovered Bou'Kongo, an early nineteenth-century settlement of slaves freed in the Middle Passage by the British navy. Burton Sankeralli, a public intellectual attached to the University of the West Indies, accompanied the expedition. Later, he published his diary. "An African liberated village," he narrates breathlessly, "the Congo nation, the cradle of the Chatham community. In the bush . . . our history in the bush . . . not quite lost" (Sankeralli 2009, 163, ellipses in original). In the same volume, Sankeralli explains, "This African village sources the ongoing living presence and soul. It provides a grounding for the ongoing living Spirit of struggle" (2009, 63). That struggle—in which Sankeralli became a leading voice—encompassed much more than the environment. "My notion of rights is tied with the notion of community," he explained to me when we met for lunch in Port of Spain in 2009.[26] He practiced Orisha, an African-derived religion, and was writing a dissertation on it. For Sankeralli, Bou'Kongo constituted a sacred site, a place where Africans had disembarked as free. Only blasphemers would bulldoze it for aluminum.

The most public opponent of the smelter, Wayne Kublalsingh, situated aluminum in a similarly historical and cosmic drama. Prior to entering politics, he had written children's books and occasionally lectured at the University of the West Indies. Then, although a slight man, he had undertaken a one-person hunger strike against the smelter. Kublalsingh moved with grace and spoke with a quiet calm that inspired the other activists around him. "He is doing this from something in his core," a follower told me, "motivated from some deep sense of connection to the land . . . and people being able to reap healthy living off the land."[27] When we met at his home in Central Trinidad, Kublalsingh began with first principles: "Spanish conquistadors smashed the Amerindian culture. . . . We've always been small and vulnerable."[28] The *we* referred to any and all Trinis, past and present. A week later, at a strategy session, he predicted a rocky road to legal reform: "over the past five hundred years of our history . . . constitutions have been implemented . . . through vast violence and blood."[29] The future might repeat the past. So it appears in the blockbuster film *Avatar* (2009). Kublalsingh watched the movie—wherein humanoids on planet Pandora eject an American mining corporation—and it reminded him of Trinidad's struggle. In a newspaper editorial titled "The Avatar Threat," he warned readers of "the corporation imperialist's [*sic*] wars . . . which natives, some in Chatham and La Brea . . . and certainly all natives on the planet, must confront as this century wears on."[30] Only Kublalsingh could stretch Caribbean history and identity to this extent and get away with it. He and his followers descended almost entirely from involuntary migrants rather than from Amerindians. Yet—compared to aluminum—even transplants to Trinidad's social landscape appeared authentic. Cans and foil glinted menacingly—at once hypermodern and bloodthirsty. Kublalsingh (2009, 72) emphasized the use of this light, strong material in advanced weaponry. Once called a "magic metal," aluminum exceeded the limits of even the most flexible pastoral sensibility (Sheller 2014, 27).

Rhetoric that worked in Chatham, however, gained less purchase in La Brea. In 2007, the government relinquished its first proposal and recommended a smelter for the site already deforested. In that depressed town, the smelter's promise of jobs garnered distinct support, creating awkward conditions for La Brea Concerned Citizens United. At the national level, this change of setting challenged the newly formed Rights Action Group. Living in the comparative comfort of Port of Spain and its suburbs, these activists found it difficult to oppose industry outright. Indeed, when I met Kublalsingh and his allies in late 2009, they refused to identify themselves as environmentalists at all. The problem, they insisted, was the size of a given industry. The footprint of development projects, they assumed or asserted, should be commensurate with the given landmass. In 2008, the economist Denis Pantin had specified this principle in a number of widely circulated essays. His text on "mega-projects in small places" recommended that planners apply an "irreversibility principle." "Given our small island reality," he wrote, "if we make an error, there will be little or no room for correction. The nuclear accident at Chernobyl affected a geographical area, for example, several times the size of Trinidad!" (Pantin 2008, 2). I met with Pantin in early 2010 in his university office and suggested that all places were small and fragile. No, he argued, larger nations are "in some way able to pick up the slack in terms of providing alternative space." "If you have one wetland," he further illustrated, "it's a different thing [than] if you have a thousand."[31] In other words, parts of continents were interchangeable and replaceable. Islands were unique. The activists' pastoralism pivoted from the agrarian to the merely small.

At exactly this time—as the smelter suffered insults in the press and at public rallies—the government raised the stakes on megaprojects. It announced an infrastructure project that seemed to burst the bounds of Trinidad's scale and its history. The Rapid Rail would connect Port of Spain to San Fernando and other major cities. Eric Williams had torn up Trinidad's first rails in the 1960s, creating conditions for widespread car use and gasoline combustion. Surely, a commuter rail system would benefit the environment, not to mention decreasing traffic congestion as well.

"Stupidity. That is total stupidity!" denounced Norris Deonarine, head of the National Food Crop Farmers Association, referring to the claim that rail would lower Trinidad's carbon emissions. The way to do so, he insisted over (imported) coffee, was to become self-sufficient in food. He reminded me of the slogan "No smelter—agriculture!"[32] As fate would have it, Rapid Rail's tracks would pass through a coastal corridor of Indian small farmers, who flooded angrily into three public consultations on the project. Such an expensive project would surely encourage government corruption, many argued. Less explicitly, the memory of indentureship haunted these gatherings. At one of the consultations, Deonarine referred to himself and others as "generational farmers," who had improved the soil quality by two grades.[33] "They now fear," explained the member of parliament at another consultation, "that all their lands that their grandparents toiled very hard to give to them are now going to be taken away."[34] Rapid Rail threatened to undo all that liberation, taking farmers back to the time of Trinidad's colonial railway, "when our people had no say whatsoever."[35] That train had carried plantation-grown cane. These occasionally far-fetched associations and equations demonized the proposed train as a one-way ticket to exploitation and landlessness. The rail corridor would bifurcate and eviscerate a dozen Bou'Kongos. Pastoralism of the agrarian sort made a comeback.

But the corridor also provoked acute anxiety related to scale. At the consultations, two participants compared Trinidad to the former metropole. Trinidad was "smaller than London and smaller than Europe," asserted Stephan Kangal of the Caroni Assembly of Villages.[36] Presumably Port of Spain needed neither the Tube nor the Eurostar. A few minutes later, Anderson Wilson of Beetham Estate took the microphone. "I travel London. I travel the world," he asserted—rather improbably, given that Beetham was notoriously poor—"Trinidad is too small for this thing."[37] In part, farmers were voicing a conventional not-in-my-backyard complaint. Few seemed to realize how narrow a rail corridor would be.[38] Just as important, though, opponents of Rapid Rail associated trains with a kind of futuristic infrastructure and streamlined mobility they thought foreign to Trinidad (cf. Larkin 2013, 334). The Rights Action Group joined publicly with the farmers. "We need to lessen the speed at which we are moving," Kublalsingh elucidated intently to me at a mall restaurant. "It is more of a metaphysical issue. . . . [The rapid rail] would give a metropolitan feel."[39] I felt that way about the eight-lane highway running by us and parallel to the proposed tracks, but Kublalsingh accepted the existing roads. Petro-pastoral sensibilities accommodated asphalt more easily than train carriages made of—or looking like—aluminum. By *metropolitan*, Kublalsingh also denoted an unwelcome, continental scale. The preferable short, insular distances, he believed, did not require a faster pace than that of the bicycle. Size—whether imagined as the footprint of the infrastructure or the way in which the technology would further compress Trinidad—seemed to trigger environmental alarms. Meanwhile, the promoters of Rapid Rail distributed and posted on the web predictions regarding cuts in carbon emissions. As motorists switched to rail, they would release 85 percent less CO_2 per passenger kilometer (Trinitrain 2010). The claim fell on deaf ears. Not a single participant in the public meetings even mentioned carbon or other forms of air pollution. Unassumingly, complicitly, the quality of the atmosphere simply mattered much less than the quantity of Trinidadian landscape.

By its very nature, this hectare-focused vigilance gave climate change, all hydrocarbons, and the specific power source of the smelter a free pass. The relevant resources lie underground and exit though tubes 30 inches wide. Of course, rigs, refineries, storage tanks, and the occasional spill enlarge this lateral spread. But, in terms of their efficient consumption of the earth's surface, hydrocarbons have no equal (Dukes 2003; Mitchell 2009, 402). Like a skyscraper, they save lateral space. At a Carnival fête, I asked Dennis Pantin about oil and gas. Did the sector constitute a megaproject and a worry for him? Not at all: "No problem," he shouted over the music, so long as rigs lay far enough apart.[40] In La Brea, the National Energy Corporation had always planned to build a 720-megawatt gas-fired power plant near the smelter so as to supply aluminum's huge need for electric current. Opponents understood this plan. Indeed, they initially spoke of a smelter complex that included a new port as well and would release multiple pollutants. Partly due to this strategic choice, the Rights Action Group never devised criticism specific to the power plant. On June 9, 2009, Gary Aboud of the organization Fishermen and Friends of the Sea issued a "national call" to demonstrate against the generator.[41] Aboud's electronic broadsheet railed against the interrelated waste of electricity, money, and natural gas. Regarding gas, the rhetoric of the Rights Action Group dwelled on the shortage of supply, rather than on the contribution to climate change. Finally, however, another member of the Rights Action Group did address

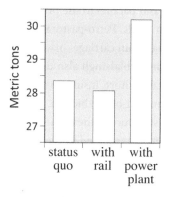

4.5 Trinidad and Tobago's options for per capita emissions, 2010. Prepared by Mike Siegel of Rutgers Cartography Lab.

carbon emissions. In November 2009, at a People's Democracy rally in Port of Spain, Cathal Healy-Singh cited "the age of global warming — when humanity itself is at risk."[42] Would he, therefore, throw his support behind public transportation? No: two months later, Healy-Singh denounced the Rapid Rail as a megaproject "contaminating this tiny land mass that we reside on."[43] No one — inside or outside the group — appeared to note these contradictions. In per capita terms, Trinidadians stood as fourth highest emitters of carbon dioxide in the world. Rail would cut Trinidad's national emissions by 1 percent and the power plant would raise them by 6 percent (figure 4.5).[44] Trinidadians ultimately chose the least sustainable option.

As it finally went down in defeat, aluminum production completely overshadowed and displaced the issue of carbon dioxide. The Rights Action Group contributed substantially to a wave of disgust with Manning's corruption and arrogance. Thinking he would win and silence his critics, the prime minister called an early election in May 2010. The smelter — but not the power plant — immediately became a focus of debate. I attended a candidates' forum in my neighborhood and asked through the moderator, "How do you assess Trinidad and Tobago's responsibility for climate change and for the reduction of carbon emissions?" "Very poorly," shot back the main opposition candidate for parliament, Annabelle Davis. The moderator resumed reading my question: "And how would you suggest that the country fulfill that responsibility?" "No smelter!" Davis retorted to hearty applause. "Simple!"[45] But it was not so simple. Davis's People's Partnership ousted the People's National Movement and shelved the smelter as well as the Rapid Rail. The government continued building the gas-fired generator at La Brea, this time with Kublalsingh's blessing. "Keep the power plant, keep the port, stop the smelter," he urged in the newspaper.[46] At a conference in early 2011, I asked Kublalsingh about the gas-fired power plant, then nearly complete. "I'm not worried about pollution there at all," he assured me. "It can be mitigated."[47] He seemed to forget about carbon dioxide. Kublalsingh and Sankeralli were in the audience when I presented this work to the University of the West Indies, mentioning the 6 percent figure. Afterward, I asked Sankeralli what he thought of that outcome. He looked at the ground, still digesting the information. Then he answered me with a pithy, less-than-scholarly expression of regret.

Sankeralli's "Oh, fuck"—to quote the unquotable—represented a breakthrough in my efforts to practice engaged ethnography.[48] I had spent an awkward year in communication with the smelter's opponents. They suspected me of supporting aluminum, even of working for the U.S. Central Intelligence Agency. I repeated continually that I opposed the smelter but perhaps not for the same reasons as every one of them. This declaration of partial solidarity could have initiated a far-reaching dialogue. How do you weigh the risks of global climate change against those of purely local import? I asked—although rarely so directly. (One learns more from extended listening and observation than from blunt interrogation.) I tried to be an engaged or activist ethnographer, that is, one who "collaborates with an organized group in struggle for social justice."[49] La Brea Concerned Citizens United and the Rights Action Group were certainly struggling for justice: they fought against the imposition of environmental risk upon Trinidad's underclass without a democratic endorsement from below (Hosein 2007). Conscience mostly ended there. I observed in meetings, speeches, and declarations a consistent neglect of hydrocarbons, carbon emissions, climate change, and all of their unjust effects. Smelting aluminum does not produce carbon dioxide. Only the source of electricity for smelting would do so. Indeed, when driven by hydrocarbons, that extraordinarily energy-intensive process releases an average of 13 tons of carbon dioxide for each ton of aluminum (Sheller 2014, 19). The gas-fired power plant, therefore, presented activists with an opportunity to protest and restrain Trinidad's carbon emissions. But they grew strangely silent. My informants' "sense of place" overrode their "sense of planet" (Heise 2008). By defeating rail and tolerating gas, they opted for the highest of three emissions scenarios. Even in opposing aluminum alone, Kublalsingh and his allies strove to relocate—rather than abolish—toxic smelting. Another island or continent now endures the risk of producing the aluminum that La Brea would have manufactured. On the radio, Healy-Singh cited the interest of "anthropologists" as evidence of the smelter's severity.[50] Until Sankeralli expressed his shock, though, activists had not valued my perspective.

La Brea's own protesters had long shielded the oil industry from criticism. Lakeside, pastoral aesthetics occluded both environmental facts and possible political strategies. Wells had rendered their community a toxic brownfield. In the United States, residents might have mobilized to demand the remediation of contaminated water and soil. They might have further claimed rights to restitution for cancers and other illnesses linked to industrial plants (Bullard 1990). Or, like those surrounding Shell's refinery outside Buenos Aires, they could have waited. Javier Auyero and Débora Swistun (2009) describe a limbo in which victims hope alternately for health or for provable, actionable pathologies. Residents of La Brea, by contrast, considered oil, gas, and bitumen as neither a threat nor a mystery. Bitumen cures the sick after all. Other hydrocarbons—and the machinery for their production—formed part of the surface topography of houses, gardens, forests, and ponds. The lake land encompassed water, petro-aqueous mixtures, and—in the case of pitch—nearly pure hydrocarbon. La Brea's flexible, even contradictory, pastoral approximated the antiurban sentiment of the Martiniquais writer Patrick Chamoiseau. "Texaco," he narrates in the novel bearing that title, "was what the city conserved of the humanity of the countryside."[51] Texaco is, in fact, a shantytown at the site of that company's former refinery in Fort-de-France. The first squatters notice the smell and danger of gasoline, but Chamoiseau emphasizes hazards from across the harbor: "The city stutters pollution and insecurity.... It threatens cultures and diversity like a global virus."[52] For La Brea, aluminum and other megaprojects posed this sort of metropolitan threat. Ultimately, smelting may have made its opponents appreciate hydrocarbons all the more, as both more humane and more natural than the alternative.

At root, these politics of place do not serve the unfolding battle against carbon-intensive development. Hydrocarbons often enjoy the most support precisely at the point of their production. There they benefit not only from pastoralism but from the entire sentiment of local belonging and history. Amid pollution, residents stretch agrarian language to accommodate hydrocarbons. The cornfield becomes an oil field. Perhaps the experience of La Brea demonstrates again the need for an ecological politics of the entire planet. Of course, this kind of global reach—the sleight of hand whereby American and other environmentalists claim authority over distant elsewheres—recapitulates numerous colonial missions (Shiva 1992). Still, the sensibility of an interconnected, interpolluting globe would add much to Trinidad's environmental debates. Consider Julian Kenny's shift in perspective. As a multidisciplinary naturalist, he chaired Trinidad and Tobago's Environmental Management Authority during the height of the smelter debate. Activists faulted him for not opposing the project publicly, but his administrative role constrained him. In that period, he wrote an essay titled "Alarmism in Science." "We seem to be too much distracted by climate change hype," he opined, "and not concentrating on the mess we are creating of our immediate environment" (Kenny 2011, 237). Widely respected but considered mercurial, Kenny refused to be interviewed. In 2011 at the Green Business Forum, I caught him off guard. "What did you really think of the smelter?" I asked. I expected him to talk about "our immediate environment." Instead, he shot back, "A couple hundred thousand tons of CO_2. . . . That was my first concern."[53] Kenny, then eighty-one, died later that year, and Kublalsingh lionized him as "an ecological messiah." Even in tiny Trinidad, perhaps, a consensus about the planet may soon supplement the obsession with place. Or we all might consider the whole Earth a place, bounded by the 20-kilometer depth of its atmosphere. Lying off the sun's shore, our island is small and, more and more, it needs a large vision.

CHAPTER 5

Climate Change and the Victim Slot

In between forays in the oil belt and conferences with oilmen, I conducted ethnography within Port of Spain's "climate intelligentsia." I apply this term to a loosely linked group of professionally successful men and women, born in Trinidad and belonging to African and Indian ethnicities. All had earned bachelor's degrees, and many had studied further in the United States, Canada, or Britain. They knew the facts of climate change, and they cared enough to join public discussions about it. To these scientists, activists, policy makers, and energy specialists, I introduced myself as a fellow traveler: an environmental anthropologist writing a book on energy policy. Together, in 2010, we participated in a round of public consultations on the country's first policy regarding climate change. The participants might have considered carbon emissions and means of reducing them. Instead, the consultations and the policy centered on impacts: environmental hazards, including even threats to oil's infrastructure. In a fashion I had not anticipated, my informants positioned Trinidad and Tobago as a victim of climate change. Evidence suggested otherwise. The country was enjoying the status of a middle-income country, with gasoline and electricity so amply subsidized that many people consumed them wastefully. Therefore, Trinidad and Tobago's per capita carbon emissions ranked fourth among nation-states (International Energy Agency 2010, 95–97). These statistics omit the oil and gas Trinidad extracts for exports. Among hydrocarbon producers, Trinidad and Tobago occupies thirty-eighth place—not an enormous contributor, but still larger than Bahrain and Ecuador combined (United Nations Statistics Division 2009, 40–72). In short, Trinidadians were collectively benefiting from the lethal hydrocarbon system and, in so doing, exacerbating climate change. Their seas rise in what Ulrich Beck (1992, 23) calls the "boomerang effect"—where pollution bounces back onto the polluter. With these informants, my conversations sometimes bordered on arguments, as instructive as they were contentious. No one broke off contact, and all seemed to consider our debate one worth having. I kept probing for an answer to the question: how and why did the climate intelligentsia frame the country as unequivocally innocent? Innocence, after all, amounts to a license to pollute.

Blame often travels in the simplest form possible—or so cultural expressions would suggest. Inhabitants of the Torres Straits, for instance, are "sinking without a trace [as] Australia's climate change victims" ("Sinking without a Trace" 2008). Victim serves as an absolute category of people both vulnerable to and innocent of the given crime. It was not always so: psychology of the 1950s and 1960s diagnosed individuals as enabling cruelty (Fassin 2009, 122). Some still blame attractive women for rape. Experts on climate change have never dabbled in this kind of ambiguity. In any case, they would have a hard time blaming the isles of the Straits: their carbon emissions barely surpass zero. But the category of victimhood has expanded well beyond the shores of this and other subsistence-level archipelagos. In the media, fully industrialized societies—ranging from China to Bahrain to Louisiana—represent themselves as victims. Hurricane Sandy swept through the energy-intensive suburbs of my state, New Jersey, leaving millions of victims but no one willing publicly to accept partial responsibility. Under new climates, hardship redeems in an almost Christian fashion. It renders or maintains the polluter's conscience pure. In this widely distributed form, I argue, victimhood increasingly constitutes a slot. Michel-Rolph Trouillot (1991) defines this term as an enduring category of thought and enquiry, one that canalizes and disciplines scholarly work. Renaissance Europe created the "savage slot," he writes, and anthropologists still explain the Other within its confines. Tania Li (2000) uses "slot" slightly differently: as a durable political tool that marks and separates "tribal" people from populations nearby and straddling the boundary. The victim slot exhibits all these features. It draws strength from archaic geographies and cleaves social groups radically and irreversibly from close comparators. Under climate change, emitters of carbon dioxide—even high emitters—have deliberately occupied or accidentally fallen into this compartment. Like the savage slot or the tribal slot, the victim slot artificially clarifies an inherently murky moral situation. It whitewashes—as innocent—societies, firms, and industrial sectors otherwise clearly complicit with carbon emissions and climate change. To the extent that the slot persuades us, it allows good people to do bad things to the biosphere. In short, the victim slot disguises complicity and displaces conscience.

In the context of climate change, innocence refers to geography as much as morality. Consider the movement for international climate justice. Using cardinal points as a shorthand, activists are pursuing a claim of the Global South against the Global North. In the course of industrializing, the North has polluted the biosphere, to the detriment of everyone but particularly to the detriment of the resource-dependent societies of the South (Davis 2010, 37; Robert and Parks 2007). In essence, Africa, Asia, the Pacific islands, and Latin America are suffering from a problem not of their own making, and they deserve various forms of compensation. This argument gives specific weight to geography and only general weight to actions. As all parties acknowledge, the North did not initially embark on this energy-intensive development pathway knowing or intending the harm. The South might well have taken the same route if it had access to equivalent finance and resources. In fact, China's rapidly expanding carbon footprint suggests an almost irresistible attraction to coal and crude. Activism centered on cardinal points then blames people as much for accidents of temperate-zone birth as for deliberate actions. The same logic exonerates residents of the tropics—people in the right place at the right time. Of course, more fine-grained analyses do break apart the reductive binary of North and South. Shoibal Chakravarty and his coauthors refer to one billion "high emitting individuals who are present in all countries" (2010, 11884). In an era of widespread neoliberalism, one might expect this citizen-centered analysis to take hold. It "responsibilizes" the consumer for his or her own choices (Goldstein 2005, 39). Yet international negotiations and policies continue to denote entire countries or societies as high or low emitting. The victim slot, in short, encompasses low-latitude land masses and especially their offshore archipelagos. Small and windblown, these islands now represent the frailty of victimhood more compellingly than does any other geography (Lazrus 2012). How did Trinidad and Trinis gain the enviable position of insular victims? Historical accident contributed as much as did strategic choices.

Islanders became victims as islands became insular. Sea-girt land forms loomed large, sometimes larger than continents in the geographical imagination of 1500–1800. Richard Hakluyt—who chronicled Sir Walter Raleigh's sixteenth-century quest for El Dorado—denoted the West Indies as "a large and fruitfull continent" (quoted in Lewis and Wigen 1997, 29). Indeed, Raleigh and other seafarers constantly sought islands as way stations that would allow them to cross water. According to historian John Gillis, an "Atlantic Oceania" of the Azores, Antillea, Atlantis, and other unverified, shifting isles connected Europe to the Indies (Crone 1938; Gillis 2004, 86). By 1800, however, new technology disenchanted islands, establishing both their actuality and their location. Isles lost their allure, and continents gained in importance. In the nineteenth and twentieth centuries, surveying, settlement, and the entire colonial project prioritized prairie, savanna, and other large expanses found only on continents. Islands even lost their function as refreshment stations: coal-fired steamers sped directly across the Atlantic. Once a patchy Pangea, small islands became wayward dots—"islanded" in Gillis's (2004) language. Demography worked against them too. In the Caribbean, in particular, total or near-total genocides almost wiped out islanders. "To the admirers of remote island peoples," writes Gillis, "innocence made them seem like the children to which they were frequently compared, vulnerable to the point of extinction" (2004, 115).

This islanding—and its attendant innocence—did not initially affect Trinidad. Gumilla treated the land mass as part of "Orinoquia," the region of Orinoco River and delta. Governor Chacón and his predecessors reported to Caracas, on the Spanish mainland (chapter 1). The British seizure in 1797 might have isolated Trinidad conceptually as well as politically. At almost the same moment, however, the epic biogeographer Alexander von Humboldt began his five-year trek through South America. In 1799, von Humboldt arrived at the T-shaped peninsula of Cariaco. There, the Gulf of Paria separates this Venezuelan appendage from Trinidad in a fashion that, for von Humboldt, called for geohistorical explanation. The gulf, he wrote, "owes its origin to subsidence and rents caused by earthquakes."[1] Humboldt believed in a dynamic, visibly fluctuating earth—and also in the reigning theory of oceanic retreat (Rudwick 2008, 106). This larger trend would soon desiccate the Gulf of Paria: "Under the actual state of things," he affirmed, "we see the coastal plains growing, gaining over the sea."[2] Among literate Trinidadians, von Humboldt's history—if not his predictions—assumed the status of fact. "All Geologists who have regarded this Island," wrote the settler historian E. L. Joseph in 1837, "agree in pronouncing it an amputation from the neighboring Continent" (1838, 4). Maps and texts of the period traced South Trinidad to the Orinoco's alluvium and mountainous North Trinidad to the Andes (de Verteuil 1858, 345; Joseph 1838, 5). In biological terms too, the island shared kin relations with South America. Had he studied Trinidad, Alfred Russell Wallace, Darwin's codiscoverer of evolution, would surely have noted the absence of any impassable Wallace Line between it and the continent proper. Species had crossed a short land bridge during the last ice age. At high water, the Amazon and Negro Rivers easily outspan the channel between Trinidad and Venezuela. In retrospect, Wallace's "Guiana District" of northeastern South America encompassed Port of Spain as well as Manaus (Quammen 1996, 74). In short, nineteenth-century speculation and observation thoroughly blurred the edge of South America.

In more practical ways too, colonial enterprises and schemes straddled the Gulf of Paria. Almost as soon as Britain took Trinidad from Spain, anti-Spanish agitators launched expeditions from its shores to liberate Venezuela. The British themselves coveted Venezuela for different reasons. In 1805, Admiral Alexander Cochrane, the father of Thomas Cochrane (see chapter 2), surveyed southern Trinidad. Surely he looked across the strait and saw Venezuela, only 11 kilometers distant. "Trinidad," he concluded, "may be said to be the key of South America, to the possession of which, the River Orinoco offers a safe and easy passage" (Cochrane 1805). Venezuela liberated and possessed itself in 1811, thwarting cross-channel imperialism. Yet in 1858, Trinidad's seminal intellectual— Louis Antoine Aimé Gaston de Verteuil—laid out the most ambitious of such plans publicized before or since. Like Joseph, de Verteuil accepted von Humboldt's fast-moving geohistory. "Even at the epoch of its discovery by Columbus," he wrote, "the Indians entertained the opinion that this catastrophe had taken place at a not very remote period" (de Verteuil 1858, 85). If floodwaters had lately isolated Trinidad, de Verteuil proposed to use them to reconnect the island to the mainland. His compendious geography of Trinidad recommended sculpting the nearby delta into a series of navigable canals. De Verteuil quoted the colonial governor's statement of ten years earlier: "Port of Spain may eventually become the receptacle of trade of that vast tract of country from which the Orinoco draws its waters" (1858, 347). Such boosterism came to naught. But it did help deny Trinidad what many other political units were acquiring at the time: a bounded "geo-body" (Thongchai 1994). Port of Spain extended to a fuzzy, indeterminate edge.

In the next century, Trinidadians—still more worldly than innocent— reoriented themselves toward other continents and islands. Although farther away, the British Empire impinged more directly upon Trinidadians than did South America. So did certain Caribbean legacies, best expressed by the first generation of black authors. No one accomplished more to name and bound a regional and race-conscious identity than did the towering intellectual C. L. R. James. Born in Tunapuna, Trinidad, almost at the turn of the century, James published his influential account of the Haitian revolution in 1938. *The Black Jacobins* metaphorically recast the region in the mold of injustice and reactions to it. "The transformation of slaves," begins James, "trembling in hundreds before a single white man, into a people able to organize themselves and defeat the most powerful European nations of their day, is one of the great epics of revolutionary struggle and achievement" (1938, ix). At that time, he envisaged independence throughout Africa and its diaspora. By 1963—as these dreams were coming to fruition—James narrowed his unit of analysis. "The history of the West Indies," he wrote in an appendix to the second edition, "is governed by two factors, the sugar plantation and negro slavery" (James 1963b, 391). Cricket was a third, more contemporary factor. As James writes in *Beyond a Boundary*, "The clash of race, caste, and class did not retard but stimulated West Indian cricket" (1963a, 72). Caste had arrived with Indian indentured workers. Of the Indo-Guyanese batsman Rohan Kanhai, James wrote, "I have found . . . a unique pointer of the West Indian quest for identity, for ways of expressing our potential bursting at every seam" (1966, 1). That regional identity seemed to inhere most in the black bowler George Constantine. His style provoked James to observe, "We West Indians are a people on our way who have not yet reached a point of rest and consolidation" (1963a, 148). Restlessly, the West Indies team beat England and dominated the world at midcentury. Through sport, James and other Trinidadians identified their island with an archipelagic team. The Orinoco's island was becoming more so.

James's student Eric Williams pushed this identity home: Trinidad as a weaker party now gaining strength. Recall that, as a midcentury historian, he sought to demolish Britain's reputation as a liberator (as explained in the introduction to this book). After 1962, as the country's first prime minister, he continued to draw attention to imperial prejudices. The rhetorical high point came in 1977 when Williams spoke at Point Lisas at the opening of the nation's first steel mill. "The colonies were to manufacture not a nail, not a horseshoe," he lectured. "They were to produce raw materials only" (Williams 1981, 82–83). That dictum had persisted through sugar into the age of oil up to the present rupture. At the Point Lisas industrial estate, Trinidad would at last harness the energy of hydrocarbons to make steel and aluminum, the latter eventually at La Brea's ill-fated smelter (see chapter 4). In a promise kept, the industrial site would have converted petroleum into downstream plastics. "Point Lisas," Williams boasted in 1977, "is the symbol also of the aspirations of the developing countries of this world" (1981, 82–83). More measurably, Point Lisas became an enormous point source for carbon dioxide. In this sense, Williams's speech may mark Trinidad's first exploitation of the victim slot. The prime minister represented heavy industry unapologetically as a right due to the downtrodden.

Beyond economic policy, geographical themes of fragility, flimsiness, and islandness have arisen periodically in public culture. During and after Williams's rule, the island's two Nobel laureates—V. S. Naipaul and Derek Walcott—waged a literary dispute centered on size, among other issues. Naipaul hardly refers to his home country without disparaging its scale. Born to Indo-Trinidadian parents, he moved to England in 1950, a teenage novelist. At the invitation of Eric Williams, he returned to write his first travelogue. *The Middle Passage* (Naipaul 1962)—whose very title seemed to relativize slavery—still angers Trinidadians. "Nothing was created in the British West Indies," opines Naipaul, "no civilization. . . . There were only plantations, prosperity, decline, and neglect. The size of the islands called for nothing else" (1962, 27). "It was hard to attach something as grand as history to our island," he recalls in a memoir (Naipaul 1988, 143). A second memoir contrasts Trinidad's "small-island geography" with the "continental scale" of Venezuela (Naipaul 1994, 214). Naipaul once joked, "Trinidad was detached from Venezuela. This is a geographical absurdity. It might be reconsidered" (1970, 34). Against this belittling of the Antilles, Derek Walcott has waged a decades-long campaign. In accepting the Nobel, for instance, Walcott reinflated his homeland in space and time: "There is a territory wider than this—wider than the limits made by the map of an island—which is the illimitable sea and what it remembers" (1992, 30; cf. Benítez-Rojo 1992). This profoundly cosmopolitan memory centers on the true Middle Passage and the voyages of Indian workers over *kala pani*, or "dark waters." His address closes with a view from Felicity, the Indo-Trinidadian heartland, imagining "the light of the hills on an island blest by obscurity, cherishing our insignificance" (Walcott 1992, 34). Trinidad, in other words, extended across oceans while oceans concealed it from view. In various ways, then, encircling water became a focal point of debate—and available for ensuing claims of victimhood.

Oases as a Diplomatic Card

At roughly the time of Walcott's Nobel award, Trinidad began to use this insular imaginary as a diplomatic trump card. In the 1990s, the country faced a choice of alliances: identify with complicit hydrocarbon producers or with the world's innocent archipelagos. Besides Bahrain, only Trinidad and Tobago—at that time—could claim belonging among both of these groups. Although it did not export enough oil to join OPEC, Trinidad did share oil and gas fields with the petro-powerhouse Venezuela. In the 1990s, it experienced a gas boom, leading to rapid capital accumulation and re-source nationalism (Mottley 2008). Why did this mineral-based pride not provoke Trinidad and Tobago's Foreign Ministry to represent the country as an oil state? Hydrocarbons never generated wealth fast enough to pro-voke an identity-shifting faith in or fear of them. Even the captains of this industry did not begin to feel secure until the gas boom of the 1990s. Port of Spain's diplomats, then, have never carried off the swagger of OPEC. In-stead, in the 1990s, they chose to huddle at the other extreme of political and economic power, with the states most prey to environmental and eco-nomic shocks. Alienated by the bluster of Tehran, Trinidad performed the suffering of Tuvalu. It joined the Alliance of Small Island States (AOSIS), a bloc that soon came to represent those most desperately vulnerable to climate change (Lazrus 2009). Indeed, this body "produced" small islands as a category and as a blameless, ethical position (Moore 2010, 116). To the main players—whom I tracked down in Trinidad years later and over the course of years—geography was the window to the soul.

In fact, Trinidad and Tobago gained admission to the club of small islands by creating it. Otherwise, its own carbon emissions might have barred Port of Spain from membership. The effort began in a hotel room in Geneva in 1990 during a meeting prior to the 1992 United Nations Conference on the Environment and Development, known as the Rio summit (Heileman 1993). Lincoln Myers, then Trinidad's minister of environment, and his two advisors agreed on a political strategy. The regional, continental blocs marginalized states of the Atlantic, Pacific, and Indian Oceans. These peripheries should unite to form a core, a strong voice. A year after the end of my main fieldwork, I returned to Trinidad, mostly so that my son could visit his friends. Perhaps because I had rented a car, dropped my son west of Port of Spain, and driven to Central Trinidad—all with breathless speed and mobility—Myers's appearance took me by surprise. He approached me in his wheelchair, vigorous but appearing meek through diminished stature. And his argument about isles matched this body language: he converted peril into moral authority. "Where else could it be," he asked, "except in an island like this—a small island like this—where all the issues concerning development and climate change can be as stark as this. . . . All the issues of development become pronounced in these finite spaces." This hazardous condition actually empowered "the smaller countries of the world." "Their resource," he continued, "the main contribution they can make, is the advocacy of justice and fair play. . . . We have to be the moral voice."[3] At diplomatic forums, at least, the meek would inherit the earth.

Leo Heileman, a marine chemist and one of Myers's advisors in Geneva, echoed this sentiment. "We didn't have economic power, political power, or military power," he recounted on a Skype line, "but we had the power of influencing the conscience of the world."[4] Weakness, it seemed, generated another kind of strength. Myers and Heileman named their thirty-eight-member group AOSIS deliberately: it sounded like *oasis*, an inverse island. (To me, Myers pronounced the acronym as *oasis*.) I met Heileman over lunch on another return visit to Trinidad. He himself was taking home leave from his post directing the United Nations Development Program in another petrostate, Equatorial Guinea. "AOSIS came out of my mind, my head," he claimed, with less meekness than Myers. Regarding Trinidad and Tobago's ambiguous position, "I had people back here bringing that point to me, raising alarms." He overrode them because, as Heileman put it again, "We [were] more placed to be the conscience of the world . . . to consider issues that are based on the environment." Conscience broke out, in other words, but only in relation to other countries' actions. Our conversation turned to Trinidad's current environmental policy. Heileman dismissed solar energy as "insignificant: . . . the scales are not there." Perhaps, his responsibilities in Equatorial Guinea—which were developing oil and gas rapidly—narrowed his sense of the possible. He advocated natural gas as a bridge fuel and dismissed as "just politics" AOSIS's current call for an 80 percent cut in carbon emissions. Equatorial Guinea, he informed me with equanimity, now sought to join AOSIS.[5]

Back in 1990, however, petrostates mostly avoided the bloc. Bahrain, whose per capita emissions stood at more than double those of Trinidad, did not join. I brought up this notable absence with Angela Cropper, the second advisor who had accompanied Myers to Geneva. She had eventually become deputy secretary general of the United Nations Environment Programme. We met in 2012 in her temporary lodgings in Port of Spain. She had taken medical leave and looked infirm. As a low-elevation island, Bahrain could have joined AOSIS, Cropper explained. But "they saw the whole climate change negotiation treaty as a potential threat." Naturally so: limits to carbon emissions might eventually dampen demand for Bahrain's oil exports. Perhaps the similarly flood-prone United Arab Emirates and mostly insular Qatar stayed away for the same reason. Why did Trinidadians—then known as the "Arabs of the Caribbean"—not appreciate their economic common interest with these Persian Gulf petrostates? Did Port of Spain anticipate switching to renewable energy? No, Cropper and her colleagues had no intention of sacrificing their country's hydrocarbon industry. They simply thought about the future only in terms of the impact—rather than the cause—of climate change. Delegates shared "the sense that all these small islands were going to be inundated. . . . [The threat] appeared more imminent than it has proved to be." In this low-grade panic, Cropper recalled, "Nobody knew where this would go. . . . The whole thing evolved really."[6] Without any conspiracy, circumstances deferred discussion of cuts to carbon emissions. Perhaps, AOSIS members were practicing what Kari Norgaard (2006, 352) calls "implicatory denial," accepting the fact of carbon emissions but avoiding the moral consequences. Or, rather, Trinidad's delegation appreciated only its own moral innocence, to the exclusion of its guilt.

After 1990, Trinidad mostly passed as a small island state in climate change's victim slot. High-placed Trinidadians didn't seem to need to perform the role. Mere discretion sufficed. Even so, at the 1992 Rio summit, the delegation found itself in an awkward position. Eden Shand, Myers's deputy, recounted the scene to me in the midst of his retirement in Delaware, from where he still ran a forestry business in Trinidad. We knew each other from my rental of his Cascade house (as described in the introduction). "They were discussing carbon pollution and pointing fingers towards the North and the Middle East," Shand recalled. "Trinidad had to be very silent 'round the table," he continued. "I remember it being an embarrassing situation." Shand winced at me from behind his beer, looking all the more pained in his stoop caused by the gravel truck on the Savannah. Amid this "strained feeling," Trinidad's delegation tiptoed through Rio.[7] Ultimately, the gathering dispelled such unease by creating a group slightly larger than AOSIS, known as Small Island Developing States. Bahrain did join this bloc (Kelman 2010, 610), and it attended the first meeting in Barbados in 1994. The resulting Barbados Declaration generously exonerated all the signatories as "among those that contribute the least to global climate change and sea level ... [while] among those that would suffer the most the adverse effects."[8] In that same year, Angela Cropper published an article titled "Small Is Vulnerable." She included no caveat for her own country. She even wrote, without qualification, "small islands because of their size are often not endowed with ... fossil fuels" (Cropper 1994, 9). As before, Cropper intended no obfuscation. Neither did an early draft of the Kyoto Protocol "reaffirming that per capita emissions in developing countries are still relatively low."[9] Trinidad and Tobago submitted that document—on behalf of AOSIS—to a 1996 preparatory meeting. Silence and omissions allowed accomplices to harbor among innocents in the victim slot.

Trinidad played no further prominent role in the global politics of climate change until November 2009. Concurrent with my fieldwork, Port of Spain hosted the Commonwealth Heads of Government Meeting, widely considered a dress rehearsal for the Copenhagen summit on climate change the following month. By that point, Eric Williams's predictions at Point Lisas had come true. A boom in gas production and downstream industries had advanced Trinidad and Tobago to the cusp of what the government heralded as "developed country status." The nation's per capita emissions had tripled from their 1990 levels—nearly the fastest rate of increase of any nation-state in that period. Meanwhile, in an effort to stabilize the climate, AOSIS was demanding immediate, drastic reductions in the use of fossil fuels. "1.5 to stay alive!" its publicity proclaimed, referring to their maximum acceptable temperature rise in degrees Celsius. Could Trinidad again carry out the trick of 1990, redeeming its emissions through international diplomacy? To do so, Prime Minister Patrick Manning would have to vindicate the country's hydrocarbon-fueled industrial policy. In part, he played with the numbers. "The atmosphere does not respond to per capita emissions," he repeated whenever relevant. "It only responds to absolute emissions." In aggregate, Trinidad and Tobago emitted only 0.1 percent of the global CO_2 total. Manning might have massaged the data further: Trinidad burned much of its gas to manufacture exports. Trinidad could have rejected responsibility—as China was doing—for these "off-shored" emissions.[10] Rather than proffer this rationale, Manning claimed a size-related exemption: at 1.3 million, the small national population pushed Trinidad and Tobago's per capita figure artificially high. At the Heads of Government Meeting itself—inside the ever-sumptuous Hyatt Hotel—Manning exercised his influence as chair to call on the Global North to compensate the Global South. The resulting document—the Port of Spain Climate Change Consensus—stipulated "a dedicated stream [of funds] for small island states and associated low-lying coastal states of AOSIS."[11] As before, no caveat excluded Trinidad and Tobago. Manning had maintained his country's position in the victim slot.

Among nongovernmental organizations (NGOs), public discussion on climate change threatened to burst beyond that restrictive category. In parallel with the Commonwealth summit—but at a markedly more plebeian hotel—NGOs convened the Commonwealth People's Forum. They invited Angela Cropper to give the opening address. Fiery and full of conviction, she declared the world to be "moving towards an ecological civilization." Amid loud applause, she asked those in the room to "accelerate the transition towards a low-carbon economy."[12] Emily Gaynor Dick-Forde, Trinidad's minister of planning, housing, and the environment, rose next to the podium. Two months earlier, the minister had claimed, "We emit very little." Grandly, she had also quoted the head of AOSIS as saying, "We are the conscience of the world when it comes to climate issues."[13] At the forum, however, Cropper's speech seemed to inspire a more humble tone. Dick-Forde referred to "that ecological civilization to which we are working." In cutting carbon emissions, she claimed, "We as a nation have been trying to do our part."[14] The statement contained more hope than truth, but, in any case, it implied responsibility. Had Cropper forced open a door? Manning and his ministers might actually have to discuss the country's own culpability. Perhaps Trinidad could balance within and outside the victim slot. "It is not one or the other," Cropper later told me wearily, sounding as if she felt personally the heavy load of Trinidad's emissions.[15]

Assessing Vulnerability

In discussions of climate change, the concept of vulnerability often conceals as much as it illuminates. It has become an indicator in Sally Engel Merry's terms, "creat[ing] a commensurability . . . even though the users recognize that these simplified numerical forms are superficial, often misleading, and very possibly wrong" (2011, 86–87). Although dubious, measures of vulnerability confer credibility upon the victim slot. Above all, the notion of vulnerability pushes responsibility to the margins. Often, of course, adverse circumstances do reduce one's scope for choice. People rarely desire to live in flood-prone areas. The housing market consigns the poor to riskier, cheaper areas. Meanwhile, climate change has hit colonized people like a blow to a downed boxer (Ribot 1995, 2009). In Siberia, for instance, Sakha herders are losing their livelihood as permafrost degrades into swamp (Crate 2008). Do they possess sufficient ecological knowledge and resilience to adapt? One hopes so, and the question and its terms fit the Sakha context. In a petrostate, however, resilience is not necessarily desirable. One might not hope that oil and gas industries bounce back—or "forward" in the latest lingo—from Katrina or the next Gulf hurricane (Manyena et al. 2011). At root, ExxonMobil and Siberian herders act as quite different agents in respect to climate change: the former propels its dynamics while the latter struggle to survive through it. The Sakha conduct their affairs as historical agents of the old-fashioned sort, generating events under conditions not of their own making. Drillers and pumpers, on the other hand, wield "technologies that . . . have an impact on the planet itself." A cloud of environmental guilt might settle among such "geological agents," but Category 3 winds blow it away (Chakrabarty 2009, 206–7). Of the three fields where the victim slot operates—islandness, diplomacy, and vulnerability—the last discourse is the most powerful and the most deceptive. In the discourse of vulnerability, Trinidad's oil and gas sector played the victim card to its greatest effect.

After the Commonwealth summit, climate-concerned politicians steered the country well away from any recognition of complicity. Prime Minister Manning began a national discussion on global warming. He had avoided the issue for decades. Before entering politics, Manning had worked for Texaco as a petroleum geologist. We met in his constituency office in June 2010. His party had just lost the election (see chapter 4), and— demoted to a mere MP—he had time to see me. Looking utterly dejected, the former statesman recalled a long period of ignorance regarding climate change. "At first, I ignored it," he admitted. He seemed to have educated himself on the topic mostly so as to reject Trinidad's status as a high emitter. Per capita measures, he argued, "discriminate[d] against small states." Had I misunderstood? "We are small. Remember that," Manning advised me. I returned to the issue of per capita emissions. "It's not right. It's not right," he insisted. "I fighting that!" In our conversation, he indicted China, which had just overtaken the United States to become the highest aggregate emitter. "They just spewing into the atmosphere," Manning accused, "and they don't care about anybody."[16] He slumped in his chair, aware that he possessed even less power than before. Manning did not seem to care that the average Trinidadian spewed five times as much carbon dioxide as the average Chinese or that China manufactured mostly for other countries. Shortly after the Port of Spain Consensus, Manning's government put pen to paper again. In March 2010, Dick-Forde's ministry released its "Draft Climate Change Policy." Of twenty pages of text, the document devoted merely two pages to vague means of reducing the country's carbon emissions. Indeed, Kishan Kumarsingh—the document's author, who had trained in chemistry and law—parroted the prime minister's line: "In a scientific context the atmosphere reacts only to absolute emissions and not per capita emissions."[17] The prime minister had closed all discussion of culpability—and, therefore, of conscience.

This rhetorical erasure became evident in public consultations on the climate change policy in early 2010.[18] This time, as civil servants, university lecturers, and NGO leaders flocked to a middle tier of hotels, each event

5.1 Kishan Kumarsingh, 2009. Reprinted with permission from the *Earth Negotiations Bulletin*.

began with Kumarsingh's note of alarm: "Sometimes a whole island is a coastal zone." For emphasis, he widened his eyes like a startled deer (figure 5.1). At the first consultation, in Port of Spain, comments from the floor backed Kumarsingh into a corner. Some participants, including myself, mentioned Trinidad's carbon emissions and suggested that the document include targets for cutting them. Eden Shand, who had returned to Trinidad for this meeting, agreed with me. He suggested Trinidad identify less with Tuvalu and more with Bahrain, Qatar, and Saudi Arabia. "If we admit our per capita prominence," he continued before his unconvinced audience, "we get to sit at the table with the big players."[19] Kumarsingh parried both of us with, "We have to bear in mind with regard to what you are asking a small country to do." Further discussion restored Trinidad to the victim position, but now as prey to solar and wind power. "Imagine that you get no electricity tomorrow," Kumarsingh warned, "because it is a green economy." In the event, the consultation did result in one concrete proposal regarding emissions. "We want Tobago to be a carbon-neutral destination," declared John Agard, a biologist and member of the Intergovernmental Panel on Climate Change.[20] Much less industrial and less populated than Trinidad, Tobago already bore the brand of a tropical paradise. Krishna Persad's eco-resort hugged its leeward coast (see chapter 3). Tourists burned jet fuel to get there, of course, but Tobagonian individuals and firms emitted little carbon. A good many already lived without electricity. They would sacrifice less going green. Fifty thousand Tobagonians, Agard implied, could more easily shoulder a burden that 1.25 million Trinidadians were too vulnerable to bear.

At a different venue, Agard almost—but not quite—dislodged Trinidad from the victim slot. In January 2010, we met in his office at the University of the West Indies. He was preparing for the climate policy consultations and had met recently with Patrick Manning. The two had debated the salience of per capita emissions. Manning, of course, cared only about Trinidad's low aggregate pollution. "Think about what it means," Agard responded, "to be a contributor to a problem of which you are also a victim. . . . Forget about the arithmetic!"[21] Nowhere else had I encountered such a pithy and forceful summary of Trinidad's ambiguous position. Hoping for more such directness, I attended Agard's professorial inaugural lecture on campus the next month. The bulk of the talk presented four scenarios in the global approach to climate change: markets first, policy first, security first, and sustainability first.[22] The first three scenarios resulted in capitalist or authoritarian dystopias of various kinds. Sustainability first, however, would allow the world to shift from fossil fuels to renewable energy with democracy and economic well-being. "That is the vision," Agard declared, beaming at his audience.[23] What did the vision mean for Trinidad's oil and gas? I queried in the question-and-answer session. "That is easy," Agard shot back, "[because it is] a wasting resource" and will run out anyway. After the formal program, I walked forward and asked Agard if he was really advocating business as usual: that Trinidad should just use up its hydrocarbons. No, he confided, it made sense to "leave something for the future" in the ground. In that case, the finitude of Trinidad's reserves made no difference: the country would stop producing oil and gas before—not because of—exhausting supplies. Ecuador had made a similar proposal to leave oil underground (Rival 2010), but nothing in Agard's presentation suggested such deliberate forbearance as a development model. Agard had overlooked this logical extension of his own sustainability first principle. It required the country to accept responsibility rather than mere vulnerability.

Fear, however, soon overwhelmed all other sentiments. By April 2010, drought and fire were scorching the country. At the second consultation, a geologist—identifying himself as "from oil"—spluttered, "There is no one alive who can remember a dry season as dry as this one."[24] This gathering actually took place in the petroleum belt almost in the shadow of the Paria Suites' mock oil rig. After Kumarsingh's presentation, a faction, smaller and less vocal than that at the Port of Spain meeting, raised the issue of Trinidad's emissions. This time, the oil and gas sector did not wait for Kumarsingh but responded on its own behalf. Shyam Dyal from Petrotrin insisted upon business as usual: "We have to realize that Trinidad is energy-based," he reminded us. "Adaptation should be given a higher priority than mitigation," he insisted before rushing out of the meeting.[25] Dyal had, in fact, overseen a study of Petrotrin's exposure to sea level rise and extreme weather events—the only risk analysis conducted in the country. Modeling of storm surges showed "catastrophic effects to onshore operations and offshore platforms."[26] "Trinidad is a small island developing state so we are vulnerable," he had told me in his office, alongside the country's oil refinery. "We have wells that could fall into the sea."[27] In this way, encircling water generated sympathy for the very industry perpetrating climate change. Back in the second public meeting, big oil became the biggest victim to global impacts. The topic of mitigation did not arise until nearly at the end, when a man objected to the draft policy's brief mention of public transport. "All I see is Rapid Rail running through Central Trinidad and demolishing endless houses," predicted the man, having identified himself with the populist "rum shop perspective."[28] The audience saw itself as doubly vulnerable: to climate change and to sustainability. I returned dejectedly to Port of Spain by ferry, where I fought fires with Akilah Jaramogi above St. Ann's. "This is reality ah climate change," she announced, weeping, "I am exhausted. I am exhausted. I am exhausted."[29] Climate change would blight her life.

In more intimate spaces such as these, an environmental conscience at last seemed to be taking shape. Toward the end of my ethnographic year, I met Winston Rudder and Keisha Garcia of the Cropper Foundation, an NGO originally created by Angela Cropper. In public the organization had criticized the oil and gas sector only for its lack of fiscal transparency (Cropper Foundation 2008). Private—but still official—communications opened up much broader issues. Submitted to the Ministry of Planning, Housing, and the Environment, Rudder and Garcia's written comments derided the draft policy on climate change. "Does the atmosphere not respond to this?" they asked in line-by-line criticism regarding increased emissions in multiple sectors. In its authors, this sarcasm must have touched a personal nerve. Garcia's husband worked for an international gas firm, and Rudder's son had trained as a petroleum engineer. Perhaps for this reason, these two environmentalists conveyed the compromises and contradictions of ecological subjectivity with uncommon sensitivity. "We want to have our cake and eat it," said Garcia, as the three of us chatted at the foundation's office. Trinidad and Tobago, she meant, wanted to become rich without relinquishing the exemptions of a poor country. Rudder agreed but was not sure how Trinidad should adjust its deep-rooted investments. "Can we go about development," he asked, "in a way that makes sense given our [environmental] responsibility and given the fact that we live on this piece of earth . . . that has a certain capacity, that has certain natural resource wealth?"[30] The question balanced parochial and universal concerns, a love of community with an awareness of its transgressions. More than a year later—on a follow-up trip—I shared lunch with Rudder at my hotel. The new government had shelved Manning's policy on climate change. Rudder seemed even less sure than before. He described a "goodness feeling about the smell" of the country's refinery. "You don't question the oil industry," he almost commanded. And, in the midst of all this silence, "We conspire in our own demise."[31]

Faced with climate change, it was easy for islanders to sound the alarm. Rising seas threatened them immediately and visibly—and also exonerated them. Especially in a European-dominated milieu, encirclement by water suggests frailty and weakness. Atolls have lain prone before natural elements as well as total genocide, slavery, and colonialism. They can credibly pass as victims in waiting of the next great injustice. Ecology still marks them as "tropical island Edens" (Grove 1995). Mostly, then, small island states do belong in the category of climate change innocents. The Maldives recently committed to cutting its carbon emissions to zero. Except under those absolute conditions, however, some islanders surely belong in the guilty camp of high emitters. Too few acknowledge this responsibility—except perhaps on the exceptional Marshall Islands. Marshallese blame themselves for impending inundation—a consequence, they believe, of allowing the United States to explode nuclear weapons on Bikini (Rudiak-Gould 2011). Their sense of guilt exceeds, so to speak, the climatological science. None of my Trinidadian informants contested that paradigm, but almost all rejected blame either tacitly or explicitly. Instead, the climate change intelligentsia situated Trinidad in a multiplex victim slot. In considering their land mass, in performing at diplomatic forums, and in planning for hazards, these experts represented their nation and their institutions as innocent. A generous pardon, it extended all the way to the country's gas rigs and petroleum refinery. The slot "rendered technical" all the thorny questions of conscience and complicity that would otherwise arise (Li 2007). Petro-Goliath entered the slot and passed as a greenwashed David. In this sense, climate change had the misfortune of being recognized by residents of small islands.

Imagine, by contrast, what can happen once continentals—in a strong nation—recognize climate change. Franny Armstrong's (2009) film follows the reckoning of a petroleum paleontologist living in New Orleans. To Alvin DuVernay, "Oil smells so much like money it's just beautiful." Then he smells corpses rotting after Hurricane Katrina. The scales fall from his eyes. We are living, he concludes, in "the age of stupid" (the title of the documentary). The charge of stupidity overlooks much complexity, but it is not a bad place to start. This portrayal leads more rapidly to accountability than does victimhood. Trinidad's new government has asserted victimhood less vocally than did Manning's administration. At the same time, no official in Port of Spain is accepting partial responsibility for climate change. Far from it: in 2012, the Ministry of Energy was simultaneously exploring for gas and launching a program of enhanced oil recovery. Still, outside the energy sector and outside government, some Trinidadians are reconsidering their nation's complicity with climate change. In our 2012 discussion, Cropper turned her earlier assumption about insularity on its head. She referred to Trinidad and Tobago as "this tiny country—which lends itself so well as a crucible for getting things done." One of those "things" could be a postcarbon society.[32] Trinidad's small size might allow it to overcome the indecision endemic to larger polities. Perhaps the proximity of everything in Trinidad throws hydrocarbons into stark relief. One can actually smell them. Perhaps, Trinidadians might appreciate the connection between hydrocarbons and sea level if they considered only the place, rather than the planet. They might understand climate change as the boomerang of their own pollution rather than as a harpoon thrown from another hemisphere. An awareness of such self-destruction might form the core of a new CO_2-specific consciousness. With luck, Port of Spain and New Orleans will assemble and export a product too rare to have a recognized name: carbon conscience.

I was living in Port of Spain when the Deepwater Horizon oil platform exploded and sank in the Gulf of Mexico. British Petroleum (BP) had drilled into the Macondo field under 5,000 feet of ocean water and through 13,000 feet of rock. Geologists and engineers had joined the heroic effort to find oil in ever-more difficult and dangerous circumstances. On April 20, 2010, gas surged up the well under high pressure. The blowout preventer failed, and the blowout killed eleven workers (Konrad and Shroder 2011). My Trini informants sympathized immediately with the dead, men largely forgotten in the frenzy of American reporting. Then, these experts criticized BP: it operated in a slipshod, unprofessional manner, lining its well with inferior cement. A Trinidadian firm manufactured better cement, and even BPTT—the local subsidiary of BP—would not have made such stupid, irresponsible errors. Safety started to sound self-righteous. As the well bled oil in its second month, I visited the office of BPTT. Just to enter the building, I had to endure a fifteen-minute safety video—mostly about where to flee in case of fire. I wondered when the industry would look up from local flames to see the spill everywhere. For the geologist Rick Bass, the Macondo well served as a teachable moment. In a new foreword to *Oil Notes*—written in the midst of the spill—he calls for "a truer accounting of the full costs of dirty carbon" (Bass 2012, xix). At about that time, however, as the spill entered its third month, my informants began to rekindle, in themselves, Bass's original enthusiasm for oil exploration. "Now do you get it?" they asked me. British Petroleum had done nothing but perforate the caprock, and geological pressure was producing huge volumes every day. This is how it comes up, they explained. It seemed beautiful, natural, and inevitable. No one said as much, but the hemorrhage at the bottom of the sea seemed to prove that oil should come up, not that it shouldn't.

How does an anthropologist position himself in the midst of such harm and such harmful thinking? When burned in large volumes, hydrocarbons wreak havoc. One cannot think otherwise without denying the findings of the Intergovernmental Panel on Climate Change. Here science and ethnography stand at cross-purposes. The ethnographer frequently searches for the common decency and goodwill that binds informants, readers, and the ethnographer himself. This thread does the work of translation, rendering the unfamiliar somewhat familiar. As a literary theme, hydrocarbons could do this job: they circulate nearly everywhere. Bass embraces his audience when he declares, "We are all complicit: the oil finders and the oil users" (2012, xix). I could have written that sort of book. But another principle of ethnography compels me to describe difference. The oil finders differ fundamentally from the oil users, a billion of whom consume next to zero anyway (Malm and Hornborg 2014, 65). Even heavy consumers driving American roads relate but distantly to the substance. Many could switch to other power sources and other technologies: buses, bicycles, or cars running on electricity generated from sunlight and wind. "Finders" work precisely to delay that substitution. They prove up supplies even as proven reserves greatly exceed what the atmosphere can safely absorb before 2050 (McGlade and Elkins 2015). These petroleum professionals live from oil, and the most passionate live for oil as well. No ethical choice would be easy for them. Most fail even to see the essential ethical choice. Christine Bader, for instance, identifies herself as a "corporate idealist." In the early 2000s, she started BP's program for social responsibility, emphasizing the rights of oil workers and neighboring communities. The spill "broke her heart." Corporate idealists, she concluded, should ask, "What are the greatest tensions that the core business of this company and industry have [*sic*] with the best interests of society?" (Bader 2014, 128, 193). Those tensions, Bader believes, center on mishaps or malfeasance at the point of production. She has only scratched the surface. Canada's industry-created Ethical Oil campaign suffers from a similar shallowness. The core business of any oil company damages the whole world. Conscience cannot abide the spill everywhere.

Near Misses

Fossil fuels were never foreordained. Near misses and contingencies have pushed Trinidad and much of the world toward hydrocarbons. Yet the most sweeping accounts of energy transitions suggest an unstoppable juggernaut. Vaclav Smil (2008, 380) refers to a "law of maximized energy flows" under which civilizations continually exploit denser fuels in more efficient ways. Nuclear fission and the latest experiments in fusion, argues the geographer Alfred Crosby, "count as triumphs in the quest of the children of the sun for more energy" (2006, 5). Perhaps the notion of a quest confers nobility on something ultimately squalid, reframing missteps as breakthroughs. Even critics—who wish to derail the train of fossil fuels— trace environmental ruin to the DNA of our species. *Homo erectus* walked resolutely out of Africa, recalls Elizabeth Kolbert, a leading popularizer of climate science. Modern humans settled the world and burned its forests and much else as well. "And now we go to Mars. We never stop" (Kolbert 2014, 251).[1] So far neither Kolbert nor the paleoanthropologist she quotes has left Earth, and no one lives on Mars. The possible technology, in other words, only becomes real under the right circumstances. Meanwhile, other possibilities bear no fruit at all. In Trinidad, chance favored oil and gas, fuels that perform far worse—in environmental terms—than the alternatives.

The island thus missed moments and movements that were both solar and utopian. In part, sunshine lacked competent champions. In 1732, Joseph Gumilla noticed a sunlit floral feast, harvested effortlessly by Amerindians and equally available to Spanish farmers. Cultivators of cacao would have to immigrate. No ship could load insolation and carry it across the Atlantic. That tether to place made solar energy more democratic. Elites could only monopolize it by monopolizing the land—a common occurrence now but less feasible in the eighteenth-century Americas. If settlers had come, they might have proved Gumilla right. Madrid, however, did not take the Jesuit seriously enough to fund his idea or even to value local lifeways. Very likely, his spellbound demeanor—inspired by the enchantment of energy and nature—failed to impress those who allocated vessels and supplies. This conundrum accompanied solar power: wonder at its unseen plenty discouraged the quantitative and managerial approach necessary to exploit it. Certainly, Conrad Stollmeyer failed to square this circle. Still, in 1845, he and Adolphus Etzler got farther than Gumilla. They recruited and transported Englishmen to a utopian colony to be powered by sun, wind, and tropical nature in general. Tropical pathogens slew the settlers before Stollmeyer and Etzler could build a converter of solar into mechanical power. In fact, the two men barely grasped the design specifications of their Satellite. In 1861, the French mathematician Augustin Mouchot patented the first solar-powered pump (Butti and Perlin 1980, 67). But, by that time, Stollmeyer was working—very competently now—with hydrocarbons. For Trinidad, the sun rose, so to speak, just a little too late.

Timing also failed in the case of somatic power. At the end of the eighteenth century, Josef Chacón knew how to harness the energy of muscle and bone, and he did harness it. He transported "arms" from other islands to Trinidad and across the wide Atlantic. Plantation slaves flowed like fuel—indeed, as the first transoceanic global fuel commodity. There was nothing utopian about this arrangement: elites monopolized the trade and, through their racism, monopolized the very idea of humanity. If solar power opened one's vision, slavery narrowed it to a thin slit. And slavery contaminated the very idea of harnessing human energy. In the nineteenth century, Trinidad tacked from one extreme to another: from the utter exploitation of human energy to revulsion at the mere hint of it. Earl Lovelace begins *The Dragon Can't Dance*—arguably Trinidad's national novel—with a reminiscence of Laventille, Port of Spain's slum. The residents' ancestors "took a stand in the very guts of the slave plantation, among tobacco and coffee and cotton and canes, asserting their humanness in the most wonderful acts of sabotage they could imagine and perform, making a religion of laziness and neglect and stupidity and waste. . . . After Emancipation . . . they turned up this hill to pitch camp here on the eyebrow of the enemy, to cultivate again with no less fervor the religion with its Trinity of Idleness, Laziness, and Waste" (Lovelace 1979, 2–3). Anyone anywhere may enjoy leisure. But it may be particularly difficult in Trinidad, the United States, and other postemancipation societies to propose muscle as a performer of work. At one of the policy consultations on climate change (see chapter 5), I recommended tree-lined bicycle lanes in Port of Spain. From Laventille or from my own neighborhood of Cascade, I suggested further, one could pedal to work in the cool shade, free of traffic and parking problems. "But I don't want that," wailed one consultant. His response seemed natural, and no Trini environmentalist challenged it.[2] Through its overreach, slavery exalted idleness and invalidated a low-carbon source of energy.

Even then—having rejected so many alternatives—Trinidad might have reaped the maximum social reward from petroleum. Agriculture and industry on the island might have used pitch, oil, and gas to underwrite postemancipation equality and leisure. Here, too, individual temperaments misaligned with technological possibility. Stollmeyer had once wished to obviate all human labor. He distilled pitch into fuel after, rather than before, his encounter with freedmen and their trinity. That experience shriveled Stollmeyer's faculties. He appears to have grown racist and even vindictive, far more eager to see blacks bent double under bitumen loads than to see them lounging under a tree. He and other capitalists deflected a potential leisure dividend toward more production. I first learned of Stollmeyer's humanitarianism through Johnny Stollmeyer, who lived along my jogging route in St. Ann's. He worked as a horticulturalist. I met him among opponents to La Brea's aluminum smelter. "We need to be preparing ourselves," he advised me, "to all live within the photosynthetic carrying capacity of our bioregion."[3] His family had certainly changed its tune, I joked. Not finding this quip funny, Johnny informed me of his great-great-grandfather's idealism. Following consciously in those footsteps, the younger Stollmeyer dreamed of small-scale agrarian villages—subsidized, presumably, by the country's wealth in hydrocarbons. Perhaps pitch and fraternal substances could, at last, pay a utopian dividend. Meanwhile, Johnny was planting trees for the liquefied natural gas plant, helping it to compensate for the destruction of landscapes in Point Fortin. Afforestation satisfied him in the short term. For one reason or another, the most free-thinking Trinis have failed to criticize the principle of burning oil and gas itself.

I lived through one of the more evident missed opportunities in 2010. Trinidad's antipollution movement had identified carbon dioxide as one of a number of risks. Was a movement against hydrocarbons about to begin? Activists protested the multipollutant smelter complex. Then, as Wayne Kublalsingh and others defeated the smelter itself, they acquiesced to the adjoining power plant, the complex's only emitter of carbon dioxide. Critics might have quashed both facilities. But carbon emissions did not rank high enough as a moral and environmental issue. The following year, as La Brea's power plant rose from the ground, it provoked a different kind of concern. Absent the smelter, how could the electrical grid benefit from a 50 percent boost in wattage? In 2011, a panel of the Green Business Forum considered this question. "We have a lot more power capacity than we do demand," lamented Dax Driver of the Energy Chamber.[4] Surplus electricity had already invalidated plans for a wind farm. Joth Singh, head of the Environmental Management Authority, conceded, "What I see . . . is a *percentage* of renewable energy on the grid, if it is going to happen at all."[5] No percentage will happen unless the country's environmental politics undergo a sea change. Now considered the environmental conscience of Trinidad, Kublalsingh has been protesting the route of a new highway not far from La Brea. He conducted a months-long hunger strike in 2014. From his bed, the emaciated man wrote against imperialism, capitalism, plantations, and—more diffidently—against heavy industry too. "The lands should be used," he insists, "to create an altered, *supplementing the oil and gas paradigm*, economic platform for the island and the Caribbean" (Kublalsingh 2014, 4; emphasis added). "Supplementing" is not sustainable. To mitigate climate change, Trinidad and all the petrostates will need to replace the paradigm of hydrocarbons. So far, contingencies, political will, and (mostly absent) conscience have backed Trinidad's status quo.

Against Fuel

Closer to my home, the politics of oil are changing. On the streets of Washington and New York, people are now challenging the spill everywhere with mixtures of hope, fear, and anger. In 2011, Bill McKibben launched a movement against the importation of oil derived from Alberta's so-called tar sands. A generation before, he had published *The End of Nature*, the first jeremiad against climate change for a popular American audience. "How should I cope," he asked in the book, "with the sadness of watching nature end in our lifetimes, and with the guilt of knowing that each one of us is in some measure responsible?" (McKibben 1989, xxv). This literary shift into a moral key did not inspire masses of American readers either to protest fossil fuels or to cut their own emissions. But, in the tar sands, McKibben found a stirring set of symbols: the heavy hydrocarbon—which also flows through La Brea—requires strip mining and chemical-intensive processing. Extraction has polluted the Athabasca River and sickened many First Nations people living downstream. If approved by the U.S. president, the Keystone XL pipeline to Houston could cause the same damage in the heartland of the United States—and would certainly raise carbon emissions. Through this geography, McKibben linked local spills to the global spill. In 2011, he forged a broad alliance between indigenous people and ranchers in the Great Plains and more conventional, coastal environmentalists. I too joined immediately, as did Eden Shand, Trinidad's former deputy minister of the environment, then living in Delaware. "I was at the front of the march," he related breathlessly to me on the streets of Washing-

c.1 Shand's Facebook post of February 19, 2013. He added the caption, "That's me with Bill McKibben, leader of the Climate Action rally in D.C. He's there for the children of the future."

ton, DC, in 2013, posting a photo to his Facebook page (figure c.1). (Note his stoop, damage done by the gravel truck on the Savannah.) Meanwhile, McKibben and his organization, known as 350.org, targeted all fossil fuels everywhere. In 2014, close to 400,000 of us marched in Manhattan. Finally, a popular movement against hydrocarbons was emerging in the United States. It has a long way to go. A TV news reporter captured my family at the end of the New York march. "David Hughes and teenage son Jesse drove in from New Jersey," she narrated—inaccurately—that night. The reporter had not asked us about our means of transport. She assumed that people cross distance in cars, and most of her story concerned the demonstration's disruption of city traffic.[6] What will it take to get more—and more influential—Americans and Trinis to rethink business as usual?

To start with, producers and users might rethink hydrocarbons entirely, as something more than fuel. A cultural reform—complementing the more explicitly political dissent—is long overdue. Geologists, economists, and other experts on oil and gas still propagate a myth of liquidity and inevitability. Stratigraphy is destiny, they feel, and the Earth practically ejects hydrocarbons. "That oil is coming up," Krishna Persad always assured me. Otherwise it would be "stranded," like a shipwrecked sailor on a desert island. What if we thought of oil as stranded in the fashion of nineteenth-century Africans, relieved to be left on their coast as the last slave ship sails away? With emancipation, elites turned their back on an energy source. Plenty of it still remained, and it still carried out a useful economic function. Somatic energy of course continued to power production—through wage labor—but never with the same throughput as in the body-consuming, body-killing sugar plantation. Simply put, no one legally uses people as fuel in industry anymore. Few can even imagine such a motivation, so immoral is slavery now considered. Oil might become the new slavery. At least some writers have suggested the analogy.[7] Canadian critic Andrew Nikiforuk refers to a "new servitude" in which "the values of one energy system have been neatly imposed on the other." Like masters of the Old South, high emitters consume energy profligately and mostly in the pursuit of luxuries and luxurious degrees of comfort (Nikiforuk 2012, 70). The historian Jean-François Mouhot confesses to his own participation in bondage because, as he argues, "Suffering resulting (directly) from slavery and (indirectly) from the excessive burning of fossil fuels are now morally comparable" (Mouhot 2011, 329). Perhaps the strain in this comparison will fade. Masters of oil will have to leave it in the ground, like slave masters relinquishing their human property and leaving Africans alone. People of good conscience will eventually strand conscienceless forms of energy. Oil will pass from inevitable to immoral to impossible.

This "new abolitionism" recalls the old, enchanted sensibility toward energy (Hayes 2014). How might one undo the monochromatic, flat attitude encapsulated in the idea of fuel? How might one revive the "moral panic" that accompanied movements for emancipation (Wahab 2010, 100)? Before that point, long before any pipelines were built, Chacón devised the idea of a disenchanted, rootless, ocean-crossing standard unit of energy. Unwittingly, he replaced Gumilla's full-throated adoration of God-given, plant-powering sunlight. Blessings became barrels. As is now clear, oil carries a vast negative blessing, a curse. Through combustion and conversion into carbon dioxide, hydrocarbons spread a scourge upon the face of the Earth, destroying natural and human communities. Increasingly, this almost religious, apocalyptic indictment rings true. But its less censorious inverse may catch on more quickly: imagine oil as a positive blessing, indeed, so powerful and so precious that one would want to use it sparingly, reverently. One might drive a car rarely and with immense fulfillment. Mimi Sheller proposes this approach to aluminum. Each 12-ounce can takes 3 ounces of gasoline equivalent to produce. Currently, we treat those containers as "cheap throw-away material." "We must become reenchanted," she pleads, "with the magic of aluminum's contribution to our capacity for lightness, speed, mobility, and flight but also wary of . . . environmental destruction" (Sheller 2014, 261). Moralized in this way, combustion would constitute a vice, pricking the conscience as a risky pleasure. Traders might still measure oil in barrels and transport it as a global commodity. Diamonds come in carats too, and the consumer proceeds with caution, releasing the mineral genie only when necessary or truly important. Of course, much else must happen: governments need to regulate oil, gas, and coal. They need to provide cheap, widespread public transportation. They need to convert electrical grids to wind and solar power. Overall, states need to undo the short-term, profit-driven capitalism under which so much of the world now lives (Klein 2014). Meanwhile, and in a less economic and political sense, anyone may help end domination by fossil fuels through veneration for them.

By the same token, anyone can embrace green energy through an act of imagination. Capitalism, markets, and so on hardly constrain us; for sunlight exceeds the bounds of any commodity form. Continuously, the sun sends 162,000 terawatts of energy into the atmosphere of the Earth, of which 128,000 remain in the terrestrial environment. By comparison, fossil fuels contribute less than 12 terawatts, a drop in the solar bucket.[8] We enjoy star rays everyday—and not primarily as electricity from solar panels. Michel Cazabon painted energy in two forms: the Pitch Lake in 1857 and, through his entire life, solar power. As described in a recent novel, he was constantly "trying . . . to see the light falling on bamboos" (Scott 2012, 459). Rays make art. They also enable surprisingly strategic alternatives to fossil fuels. Shortly after he joined Trinidad's Carbon Reduction Strategy Task Force, Krishna Persad invited me to a one-day cricket match. Sitting in stands named after Conrad Stollmeyer, he shared his idea of piping natural gas to every home in Trinidad. Residents would run their clothes dryers directly on natural gas, rather than less efficiently on electricity derived from gas. "I've got something better than any of that," I boasted, "a solar-powered clothes dryer." "Really?" he turned away from the game and toward me. "What's the technology?" "It's a long, thin technology," I said coyly, "fairly cheap and widely available." "It's not available here," he contradicted me. "Do you have it up in the States?" "Yes, but it works much better in Trinidad, at lower latitudes. We went around like this, slowly and somewhat stupidly because of the rum Persad had thoughtfully brought. Finally, laughing, I disclosed the technology: a clothesline. Sunlight will not be bottled—at least not nearly all of it.

Like the young Conrad Stollmeyer, I dream of a utopia. Utopias begin with a revolution in political and economic conditions and culminate in a "new person." Imre Szeman calls for "new ways of making subjects, which can be the only hope for the planet we collectively inhabit" (2014, 462). Such a reform may unfold with less effort than Szeman implies. It begins with filling the moral void around energy. In that space, high emitters would express a growing sense of responsibility for climate change. Anyone might wonder at energy. Fusing both sentiments, this new subject would subscribe to a postfuel notion of the ability to do work. In connection with wind power, for instance, Robert Righter (2002) describes "energy landscapes" pulsing with blades both beautiful and technologically sublime. Harvesting energy from the planet's surface in this way invites people to reengage with their surroundings. Neighbors of turbines see energy daily. Rather than merely consuming it by the gallon or the kilowatt, they cohabit with it. Or they collaborate even more concretely. Andrew Mathews (2014, 6) refers to "domesticating the carbon cycle" as Italian foresters gather energy from biomass. They are not merely cultivating, harvesting, or harnessing wood. On a larger scale, they understand their role in a planet-wide circulation essential to life and due for rebalancing. Thus, new thinking about energy might focus simultaneously on the near at hand and on far-reaching journeys. I do not mean to suggest only that one treat certain commodities as fetishes of good conscience (Carrier 2010). We should consume less and, first of all, notice the flow of these substances into and through our lives. Sustainability, then, benefits from attention and mindfulness to objects and the energy consumed in making them. In this form, we might find an attainable utopia: a way of treasuring the ability to do work.

At root, I am asking you to imagine what energy has lost. As the history and ethnography in these pages make clear, energy has become an object of political economy—and merely that. Readers may interpret the foregoing chapters in two ways. First, I have traced the pathways of various hydrocarbon commodities: bitumen, oil, and natural gas. In each case, supply and demand became and remained robust. Even before hydrocarbons, certain residents of the Caribbean demanded slaves and, in so doing, strung together the first intercontinental energy market. None of these protagonists, though, has simply bought and sold. They have imagined energy as one thing and not as another. Here is the second gloss on *Energy without Conscience*. From Chacón to Stollmeyer to Persad to Kublalsingh to Manning, influential Trinis have constructed a mental model of the ability to do work. As they bought, sold, and debated that good, they branded it as one thing: as a necessary, available, unquestionable means to everything modern. Even as modernity transformed one product after another—from sugar through to plastics—producers and consumers perpetuated this narrow vision of energetic means. In imagining those means as fuel, they cut off other ways of thinking about energy. Not deliberately—but systematically, nonetheless—all parties to Trinidad's oil economy exempted the substance from moral analysis. Here is the greatest complicity: the failure to consider alternatives and to apply conscience to those choices. Throughout the hydrocarbon age—in Trinidad and beyond—so many people have extracted and burned so much with so little pause or reflection. What if one did pause and consider paths not taken, options once available and perhaps still at hand? Only a handful of my informants—people like the politician-turned-protester Eden Shand—willed themselves to see the profound decision all around them. So many other people have, in a blandly unimaginative way, brought the world to the brink of disaster.

The nagging question that remains is one of attitude. In what tone—and on what common ground—should one write or speak of fossil fuels and their loyalists? What can an anthropologist and an ethnographer contribute through writing? Occasionally, in *Energy without Conscience*, I have employed the condescending, judgmental tone of one who sees the future. Perhaps I should apologize for insulting some Trinidadians, for labeling them as complicit and conscienceless in the face of planetary harm. Rather than retract, I will end more bluntly still: the petro-geologists among my informants are in the wrong and doing wrong. I did not find them to be exceptionally greedy or underhanded, but I did detect a moral problem. They take credit for producing hydrocarbons while disavowing blame for climate change. The costs of this abdication remain obscured, but soon they will break into view and provoke a widespread rejection of fossil fuels. I write with a bias for optimism—what the economist Albert Hirschman once called "a passion for the possible" (1971, 26). Others share this hope for a low-carbon future. Indeed, virtually the whole world already acts in accord with this positive scenario. Few among us are preparing in any practical way for the converse: the runaway rise in sea level and extreme weather that more hydrocarbons guarantee. Trinis are not moving from the coast. Illogically perhaps, they refuse to surrender it to the planetary depredations of their own leading industry. We are all banking on a rapid economic and political shift to sustainability. Perhaps some believe carbon capture and storage will solve the problem singlehandedly. The rest of us consign oil firms to an ash heap, worthy of condescension and worse. Perhaps this is the most hopeful finding of all: on the plane of unacknowledged assumptions, governments, firms, and individuals have already replaced coal, oil, and gas. All the dissident must do now is recognize and assert what so many assume. Any tone in any medium will help. Humor and wonder and science and art—as well as outrage and rage in the streets—will move the world to burn far less fossil fuel. Conscience will replace complicity. Obama has prohibited construction of the Keystone XL pipeline. Shand has returned to Trinidad and wishes to install wind turbines on the north coast.

Introduction

1 Edwin Drake drilled a well in 1859 in Titusville, Pennsylvania, which is often credited as the world's first. The Drake well, however, produced very little oil.

2 With a less masculinist and more nationalist cast, Machel Montano sang the lyric with David Rudder in "Oil and Music" in Montano's *Flame On* album (2008). I am grateful to Marc White for his assistance in tracing the phrase.

3 "Este libro versa sobre una novela que no existe. Y no hay en ello ninguna hipérbole. No se da en Venezuela una novelística del petróleo, como, por ejemplo, está presente en el ámbito hispanoamericano una novelística de la revolución mexicana" (Carrera [1972] 2005, 27).

4 The photographer titles this section "Oil Wells, Kuwait." Oddly, the section titled "Oil, Baku, Azerbaijan" shows pipes, valves, rigs, and no oil at all.

5 "Cracha noir. . . . C'est du charbon. . . . J'en ai dans la carcasse de quoi me chauffer jusqu'à la fin de mes jours" (Zola [1885] 1968, 37).

6 "Greffes" (Zola [1885] 1968, 133).

7 "Bête mauvaise . . . la guele toujours ouverte, qui avait englouti tant de chair humaine!" (Zola [1885] 1968, 442).

8 Appel, Mason, and Watts (2015a, 10) refer to representations that reduce oil to a "mere metonym" for modernity, money, geopolitics, violence, and commodities.

9 "Oro se engendra en tierras estériles y adonde el sol tenga fuerza." Carta-Relacion del Cuarto Viaje de Cristobal Colon, Jamaica, July 7, 1503 (Pérez de Tudela et al. 1994, vol. 3 1527; cf. Gómez 2008, 400).

10 For a more materialist reason, Sharife (2011) refers to a contemporary "colonizing" of the atmosphere through carbon pollution.

11 Crutzen and Stoermer (2000) first coined the term in a less widely circulated publication.

12 I am grateful to an anonymous reviewer for suggesting this comparison.

Chapter 1: Plantation Slaves, the First Fuel

1 From Joseph Gumilla to Governador y Capitán General, Guayana. Archivo General de Indias (AGI), Signatura Santo Domingo 632 (quoted in Gumilla 1970, xvii n3).

2   M. Roume de St. Laurent to Don Juan de Catilla, March 20, 1777 (reprinted in Borde 1882, 380–82).

3   "L'établissement de la Colonie de la Trinité et Moyens de la porter promptement à sa perfection." Philippe Rose Roume de Saint Laurent, June 2, 1783, AGI Caracas 443 (cf. Besson 2010, 34).

4   Borde (1882, 382ff.) reprints the cédula.

5   "Libertar para siempre à los Esclavos de todo derecho de Importacion, atendiendo al aumento que el ello resultaría à la agricultura." Francisco de Savaadra, June 30, 1784, AGI Caracas 443.

6   Josef Maria Chacón, Informe, November 22, 1784, AGI Caracas 444.

7   Reprinted from the *Trinidad Gazette*, June 1825, in Fraser (1971, 194).

8   Some other observers and other colonies clearly treated the labor force as a machine, a semidurable apparatus that itself consumed fuel. Slaves needed to eat. Roume, a planter himself, had articulated the need for estates to provide space for food crops. Jamaica took this thinking a step further; it imported breadfruit trees from Polynesia precisely to feed the enslaved population efficiently and cheaply (Newell 2010). In an even more machinelike way, a slave population might produce more slaves. The owners of Mount Airy, a tobacco estate in Virginia, bred humans like livestock (Dunn 2014). In effect, the female population functioned as a reproductive factory. Even when female and male slaves produced sugar, they functioned physically as "dehumanized cogs in a very inefficient machine" (Dunn 1972, 324). Through what Marx (1976, 283ff.) calls a social "metabolism," they fashioned foodstuffs and the land itself into commodities.

9   "Acclimitasés." Roume de St. Laurent to Secrétariat d'État de la Marine, February 3, 1785, Archives Nationales d'Outre-Mer, Col. C^{8A} F° 344, http://anom.archivesnationales.culture.gouv.fr/ark:/61561/zn401wqsrwwo.

10 "Ademas de estranar al clima . . . demas arboles de cavezuela menor descuida de los blancos que estan all . . . es beneno que los mata." Josef Maria Chacón to Josef de Galvez, October 8, 1784, AGI Caracas 152.

11 "Nosotros no huvieramos podido cumplir lo pactado sin grandes perdidas se ella huviera introducido en los 4 ultimos meses los 4000 Negros que ofreció." Josef Maria Chacón to Josef de Galvez (no. 8), February 22, 1785, AGI Caracas 152.

12 "Qualquier Casa de Comercio Europea que consiga vender annualmente 4 o 6000 Negros al contado al precio de 750 pesos cada uno; puede mui bien fiar el numero de 7000 Negros pagados al termino de un año." Josef Maria Chacón to Josef de Galvez (no. 8), February 22, 1785, AGI Caracas 152.

13 "Es una de las tareas que me ocupan insesantemente. Los esclavos de la parte de la costa de Africa que frequentan los Portugueses son preferibles à los que nos traen los Ingleses, Franceses, y Americanos, asi por mas dóciles, como por mas haviles y robustos." Josef Maria Chacón to Sr. Marques de Sonora (no. 87), March 10, 1787, AGI Caracas 152.

14 "Atraher Colonos que tengan Esclavos y fondos para hacer Casas, y dedicarse desde luego al cultivo de la tierra." Circular sobre Poblacion de Trinidad, January 1, 1786, AGI Caracas 444.

15 Josef Maria Chacón, November 22, 1784, AGI Caracas 444.

16 "Que reemplasan el numero de los que han muerto." Josef María Chacón to Sr. Marques de Sonora, March 10, 1787, AGI Caracas 152 and 444.

17 "Las enfermedades que causan las primeras exalaciones de unas tierras que no habian visto el Sol quizá desde que salieron de las manos del Criador." Josef María Chacón, August 18, 1792, AGI Caracas 444.

18 "La perdida que han sufrido este año por falta de Mulas con que moler las cañas y atender a los demas trabajos de sus haciendas." Josef Maria Chacón to Sr. Don Diego de Gardoqui (no. 79), June 19, 1796, AGI Caracas 153 and 444.

19 "Destruido enteramente el monton de Negros fugitivos . . . enteramente sosegada la Isla y los Esclavos con toda sugecion, y seguridad." Josef Maria Chacón to Sr. Josef de Galvez (no. 10), [month illegible] 15, 1784, AGI Caracas 152.

20 "Acantonamientos . . . propuse à su Magestad el establecimiento de una Compania de Dragones . . . suficiente para attajar y coger los Negros fugitivos." Josef Maria Chacón to Antonio Valdez, March 7, 1788, AGI Caracas 152.

21 "Los Ingeleses y Holandeces en la precision de reconocer Independentes los esclavos fugitivos, comprando à expensar de tratados bergonzosos una Paz que no havian podido conceguir con la fuerza, y en la que por consiguiente no pueden afianzar su confianza." Josef Maria Chacón to Antonio Valdez, March 7, 1788, AGI Caracas 152.

22 Josef Maria Chacón to Antonio Valdez, March 7, 1788, AGI Caracas 152.

23 "Grave daño de sus Amos, y de si mismo pues una vez sacudida la Esclavitud, andan errantes, y entregados à la desidia y al vicio, de que siguen perniciosas Conseqüencias." Francisco de Savaadra to Josef Maria Chacón, June 30, 1784, AGI Caracas 443.

24 How necessary was all this preparation? When combustible petroleum came, in the 1850s, elites would likely have embraced this dense energy package as supremely useful under any circumstances. It might have gained currency on this basis alone, unassisted by a prearranged cultural understanding. Still, the construction of other prototypes indicates a sense among planters that something new was required. The abolition of slavery in 1838 provoked a search for nonsomatic substitutes: "The great aim of the planter," wrote the French creole Louis Antoine Aimé de Verteuil in 1848, "must now be a reduction in manual labour" (de Verteuil 1848, 2). In an essay commissioned by the governor, this first geographer of Trinidad considers the problem of exhausted soils. Yields were falling on Trinidad's leading cane plantations. Before Emancipation—and especially with Chacón's incentives—planters might have dispatched slaves to cut new plantations perilously from the forest. De Verteuil does not even mention this possibility. His "scientific principles of agriculture" call for chemical means of restoring extant fields (1848, 3). "The solid and liquid excrements of

animals," de Verteuil details, "are the best manures of those plants upon which they have been fed," including cane (1848, 56). The fuel is dense. Of course, dung had long fertilized fields, but de Verteuil applied newfound quantitative principles: a horse defecates 50 pounds per day, which, when distributed at 60–75 cartloads per acre, will raise yields from as little as 1,000 pounds to as much as 2,200 pounds of sugar per acre (1848, 58–60). Here is a whole scientific system of production, distribution, and application. Although issuing from the backside of an animal—rather than the topside of a well—manure performs the same job in the same way. So does bat guano, and much of the world participated in the guano rush in the mid-nineteenth century (Hager 2008). Loosely speaking, all these fuels substituted for bonded men in the farm field. Once invented, slavery/fuel constantly reinvented itself.

Chapter 2: How Oil Missed Its Utopian Moment

1 "Le jeune Bastien, pour se reconnaître envers Celiante qui l'a obligé dans divers services, ne manquera guère de lui offrir la preuve de gratitude qu'un jeune homme de vingt ans peut offrir à une dame de cinquante" (Fourier 1840, 7).

2 C. F. Stollmeyer, "Satellite," *Morning Star*, October 11, 1845, emphasis in original.

3 "The Second Tropical Emigration Society," *Morning Star*, October 18, 1845.

4 "Tropical Emigration Society Report of the Directors, Read and Adopted at the Annual Meeting Held January 4th, 1846," *Morning Star*, January 17, 1846.

5 Excerpted in "Review," *Morning Star*, May 3, 1845.

6 "Review," *Morning Star*, June 7, 1845. Although printed entirely in quotation marks, this passage appears to have adapted, paraphrased, and expanded upon Hall (1827, 103–4).

7 "Review," *Morning Star*, June 7, 1845. Again, quotation marks indicate that Duncan attributes this passage to Hall, but Hall's text contains nothing resembling it.

8 Excerpted in "Review," *Morning Star*, July 12, 1845.

9 James Elmslie Duncan, "On Climate, Particularly of Venezuela and the Tropics," *Morning Star*, May 31, 1845.

10 From Thomas W. Carr and Charles Taylor to the Directors of the Tropical Emigration Society, October 20, 1845; printed in the *Morning Star*, November 29, 1845.

11 From Thomas Carr to C. F. Stollmeyer, March 4, 1846, printed in the *Morning Star*, April 18, 1846.

12 From W. E. Prescod to the Directors of the Tropical Emigration Society, March 7, 1846, printed in the *Morning Star*, April 18, 1846.

13 C. F. Stollmeyer to the Secretary of the Tropical Emigration Society, January 20, 1846, printed in the *Morning Star*, February 28, 1846.

14 *Gazette* (Port of Spain), May 5, 1846, reprinted in "The Tropical Emigration Society," the *Morning Star*, August 1, 1846.

15 *Gazette* (Port of Spain), May 5, 1846, reprinted in "The Tropical Emigration Society," the *Morning Star*, August 1, 1846.

16 Charles Stillwell, "On Climate; Particularly of Venezuela and the Tropics," *Morning Star*, June 14, 1845, emphasis added.

17 C. F. Stollmeyer to the Secretary of the Tropical Emigration Society, January 20, 1846, printed in the *Morning Star*, February 28, 1846.

18 *Gazette* (Port of Spain), May 5, 1846, reprinted in "The Tropical Emigration Society," the *Morning Star*, August 1, 1846.

19 From Thomas W. Carr to C. F. Stollmeyer, March 4, 1846, printed in the *Morning Star*, April 18, 1846.

20 Thomas Powell to the Editor, May 21, 1846, printed in the *Morning Star*, July 4, 1846.

21 C. F. Stollmeyer to W. E. Gladstone, Secretary of State, Port of Spain, February 20, 1846, Public Records Office, London, CO 295/151, 22, emphasis added. I am grateful to Selwyn Cudjoe for sharing this document.

22 "Numancia," *Trinidadian*, December 1, 1852, emphasis in original.

23 "Emigration to Venezuela," *Trinidadian*, November 6, 1852.

24 "A few Words on Emigration to Venezuela," *Trinidadian*, December 8, 1852.

25 "Manufacture of Fuel from Bitumen," *Trinidadian*, August 10, 1853.

26 "The Future Prospects of Trinidad," *Trinidadian*, February 16, 1853.

27 "Immigration and the Prospects of Trinidad," *Trinidadian*, January 22, 1853.

28 Conrad F. Stollmeyer to Arthur Craig, December 1855. I am grateful to Steven Stoll for sharing this document, which is in his private possession.

29 "Most Violent Assault upon the Editor of the The *Trinidadian*," *Trinidadian*, August 6, 1853.

30 C. F. Stollmeyer vs. J. Kavanaugh, Supreme Civil Court, Port of Spain, reprinted in *Port of Spain Gazette*, October 7, 1853.

31 "Most Violent Assault upon the Editor of the The *Trinidadian*," *Trinidadian*, August 6, 1853.

32 "The End," *Trinidadian*, December 24, 1853.

33 For example, *Trinidadian*, March 9, 1853.

34 Conrad F. Stollmeyer, "Raw Asphalt as Auxiliary Fuel with Megass, Wood, or Stone Coals," November 1871, Cochrane Family Papers, Box 8, Duke University Library Archive, Durham, North Carolina.

35 That country contained enough petroleum—from which kerosene could also be distilled—to serve illuminative and mechanical purposes.

36 Conrad F. Stollmeyer to the Editor, *Trinidad Chronicle*, August 7, 1866.

37 Quoted in de Verteuil (1994, 100). Anthony de Verteuil, who is a descendent of L. A. A. de Verteuil, gives no further information on the letter or on its original language.

38 Conrad F. Stollmeyer to Arthur Craig, December 1855. I thank Steven Stoll for sharing this document with me.

39 Conrad F. Stollmeyer to James McAlley, November 8, 1871, Cochrane Family Papers, Box 8, Duke University Library Archive.

40 *Creole Bitters*, May 3, 1904.

41 There are, in fact, many ways in which to calculate this figure. For a discussion, see de Sousa (2008).

42 "La machine est la rédempteur de l'humanité, le Dieu qui rachètera l'homme des sordidœ artes et du travail salarié, le Dieu qui lui donnera les loisiers et la liberté" (Lafargue [1880] 1994, 59).

Chapter 3: The Myth of Inevitability

1 The scientific term for tar sands—also known as oil sands—is bituminous sands.

2 Lately, however, the exploitation of shale gas in the United States has provoked a reevaluation.

3 The figures are 2,795 versus 565 gigatons of CO_2. The former number includes oil, gas, and coal (McKibben 2012). Bridge and Le Billon (2013, 65–66) give a figure of 620 gigatons of CO_2 for proven oil and gas reserves, still higher than the climate boundary. See McGlade and Elkins (2015) for the most thorough analysis.

4 Scott (1998) has provoked much debate on bureaucratic, improvement-oriented, and homogenizing ways of seeing. Ferguson (2005) argues that oil companies, by contrast, see territory in a way that emphasizes heterogeneity. I write of oil "producers" so as to distinguish the same actors' view of underground resources from their models of aboveground risk.

5 Coll (2012, 541) quotes an offended advisor to President Obama on energy issues.

6 I borrow the terms *traverse* and *columnar* from Rudwick (1976, 164).

7 Krishna Persad, conversation with the author, La Romaine, Trinidad, January 5, 2012.

8 Krishna Persad, "Future Hydrocarbon Prospects in Trinidad and Tobago's Explored Basins," presentation to the Energy Conference, Port of Spain, Trinidad and Tobago, February 6–8, 2012.

9 The Society of Petroleum Evaluation Engineers and the Society of Exploration Geophysicists also coauthored the document (Society of Petroleum Engineers, et al. 2011).

10 Larry McHalffey, remarks at the release of the National Gas Reserves Audit, Port of Spain, July 13, 2010. See Breglia (2013, 62–63) for a similar account from Mexico.

11 Renuka Singh, "Ten Years Left," *Express*, July 14, 2010.

12 David Renwick, conversation with the author, Port of Spain, July 12, 2010.

13 "Energy Chamber to Govt on Falling Gas Reserves: Take Action Now," *Express* (Port of Spain), July 21, 2010.

14 Philip Farfan, remarks at the Understanding Reserves workshop, Energy Conference, Port of Spain, February 8, 2012.

15 In fact, Neanderthals went extinct, except to the extent that they interbred with modern humans.

16 Farfan, conversation with the author, Port of Spain, January 7, 2013.

17 For recent mentions of this term, see Bridge and Le Billon (2013, 21) and Moors (2011, 11).

18 Prior to 2010, in fact, the group was known as the South Trinidad Chamber of Industry and Commerce.

19 Thackwray Driver, conversation with the author, New York, April 4, 2012.

20 Driver, conversation with the author, Maracas Beach, Trinidad and Tobago, January 6, 2013.

21 Krishna Persad, presentation to the workshop Business Opportunities from Green House Gas Mitigation Measures, Port of Spain, January 27, 2013.

22 Persad, conversation with the author, Port of Spain, January 27, 2013.

23 Persad, conversation with the author, La Romaine, February 24, 2010.

24 Clyde Abder, conversation with the author, St. Augustine, Trinidad and Tobago, April 26, 2010.

25 Shiraz Rajav, conversation with the author, Port of Spain, April 22, 2010.

26 Selwyn Lashley, remarks at the launch of the Carbon Reduction Strategy Task Force, Port of Spain, April 28, 2010.

27 Port of Spain, May 11, 2010.

28 Abder, conversation.

29 Persad, conversation with the author, Port of Spain, June 3, 2010; emphasis in original.

30 Carolyn Seepersad-Bachan, remarks at the Green Business Forum, Port of Spain, March 23, 2011. She did not mention Persad by name.

31 Persad, conversation with the author, San Fernando, January 5, 2012.

32 Persad, La Romaine, February 24, 2010.

33 Vincent Pereira, remarks at the Energy Conference, Port of Spain, February 6, 2012.

34 Persad, conversation with the author, Port of Spain, January 7, 2012.

35 Peter Wyant, conversation with the author, Port of Spain, January 8, 2012.

36 Persad, La Romaine, February 24, 2010.

Chapter 4: Lakeside, or the Petro-pastoral Sensibility

1 "Se hundió una mancha de tierra por donde estaba el camino, y luego en su lugar remaneció otro estanque de Brea, con espanto y temor de los vecinos, recelos de que quando menos piensen, suceda lo mismo dentro de sus Poblaciones" (Gumilla [1745] 1945, 47).

2 Arthur Forde, conversation with the author, La Brea, February 11, 2010.

3 Ethelbert Monroe, conversation with the author, La Brea, March 2, 2010.

4 Errol Jones, conversation with the author, Port of Spain, June 10, 2010.

5 Conversation with the author, La Brea, March 7, 2010. I never got his name, and, if I had, I would be using a pseudonym anyway.

6   Virginia Piper, conversation with the author, La Brea, January 28, 2010.

7   Joshua Logan, conversation with the author, La Brea, February 11, 2010.

8   Noah Premdas is a pseudonym.

9   Conversation with the author, Union, Trinidad, October 27, 2009.

10 The reading taken downstream from the reservoirs, at the mouth of the Vessigny River, showed 87.0 mg/L, as compared with Trinidad and Tobago's limit of 10 mg/L. The U.S. Environmental Protection Agency stipulates only 0.01 mg/L as the threshold for safe drinking water (Institute of Marine Affairs 2003, 41). See Agard (1988) for data regarding petroleum pollution in the Gulf of Paria.

11 Noah Premdas, conversation with the author, Union Village, October 27, 2009.

12 Isaac Gregory, conversation with the author, La Brea, October 28, 2009. Isaac Gregory is a pseudonym.

13 Adam Chalant, conversation with the author, La Brea, November 16, 2009. Adam Chalant is a pseudonym.

14 Alfred Antoine, "Why They Arrest Here," handwritten calypso lyrics, 2010.

15 Conversation with the author, La Brea, 19 May 2010.

16 Xante, "The wanderer," track 10 on "Jump Start" compact disc, no date.

17 Roger Achong, conversation with the author, La Brea, March 15, 2010.

18 Wendy Kalicharan, conversation with the author, San Fernando, January 15, 2010.

19 Ayana Kalicharan, conversation with the author, San Fernando, January 25, 2010.

20 "Ivan Kalicharan MAS 2010" [brochure], n.d.

21 Molly Gaskin, conversation with the author, Pointe-a-Pierre, January 13, 2010.

22 Reeza Mohammed, conversation with the author, Point Lisas, February 12, 2010.

23 Douglas de Freitas, conversation with the author, Freeport, April 29, 2010.

24 Conversation with the author, Port of Spain, January 25, 2010.

25 Chalant, conversation, November 16, 2009.

26 Burton Sankeralli, conversation with the author, Port of Spain, September 16, 2009.

27 Conversation with the author, Port of Spain, February 4, 2010.

28 Wayne Kublalsingh, conversation with the author, Arouca, September 18, 2009.

29 Kublalsingh, remarks at the Republic Day Conference of Civil Society Organisations, Port of Spain, September 24, 2009.

30 Wayne Kublalsingh, "The Avatar Threat to La Brea, Claxton Bay," *Trinidad and Tobago Guardian*, January 13, 2010, A25.

31 Dennis Pantin, conversation with the author, St. Augustine, December 17, 2009.

32 Norris Deonarine, conversation with the author, St. Augustine, June 16, 2010.

33 Deonarine, remarks at Trinitrain Public Consultation, Tunapuna, April 8, 2010.

34 Tim Gopeesingh, quoted in the transcript of the Trinitrain Public Consultation, Port of Spain, April 6, 2010.

35 Remarks at Trinitrain Public Consultation, Chaguanas, April 7, 2010.

36 Stephan Kangal, remarks at Trinitrain Public Consultation, Chaguanas, April 7, 2010.

37 Anderson Wilson, remarks at Trinitrain Public Consultation, Chaguanas, April 7, 2010.

38 Many may have confused the 1-kilometer-wide study area with the eventual rail corridor, sure to be a fraction of that width.

39 Kublalsingh, conversation with the author, Trincity, February 5, 2010.

40 Pantin, personal communication, St. Augustine, January 30, 2010.

41 This notice was distributed widely via e-mail.

42 Cathal Healy-Singh, speech at the People's Democracy rally, Woodford Square, Port of Spain, November 22, 2009.

43 Cathal Healy-Singh, interviewed by Gideon Hanoomansingh, *Issues and Perspectives*, Heritage Radio 101.7 FM, Port of Spain, January 19, 2010.

44 According to publicized projections, the train would reduce national CO_2 emissions by 414,000 Mt annually (Trinitrain 2010, 6), 1.1 percent of the country's emissions of 38.0 million Mt in 2008 (International Energy Agency 2008, 46). The per capita figure would fall from 28.37 to 28.06 Mt. However, the projections applied to 2032, the anticipated completion date of the entire rail system. Regarding the smelter, the corporation building it (Alutrint) projected its daily consumption of natural gas as 121 mcf (in public presentations at Vessigny on December 13 and 17 and March 5 and 11, 2007, and at Couva on December 17, 2007; Rapid Environmental Assessments 2006). At that rate, the plant would generate 2.5 million Mt of CO_2 per year, a 6.5 percent addition to the 2008 national output. The per capita figure would rise to 30.21 Mt. When completed in late 2011, the power plant was only running at 35 percent capacity (250 of 720 MW). Therefore, the increase in carbon emissions at that point amounted to 2.3 percent.

45 Annabelle Davis, remarks at Meet the Candidates forum, St. Ann's Cascade Hololo Community Group, Chinese Association, St. Ann's, May 18, 2010.

46 Wayne Kublalsingh, "The Correct Way to Stop the Smelter," *Trinidad and Tobago's Newsday*, June 20, 2010.

47 Kublalsingh, conversation with the author, St. Augustine, March 26, 2011.

48 Kublalsingh, conversation, March 26, 2011.

49 Goldstein (2012, 35), citing oral comments by Charles Hale.

50 Healy-Singh, interview by Hanoomansingh. He was speaking of myself and Simone Mangal.

51 "Texaco était ce que la ville conservait de l'humanité de la campagne" (Chamoiseau 1992, 360, my translation).

52 "La ville . . . saccade des pollutions de l'insécurité; elle . . . menace les cultures et les différences comme un virus mondial" (Chamoiseau 1992, 443–44; my translation).

53 Julian Kenny, conversation with the author, Port of Spain, March 24, 2011.

1 "Les deux golfes [Paria and Cariaco, to the west of the peninsula] doivent leur origine à des affaissemens et à des déchiremens causés par des tremblemens de terre" (Humboldt and Bonpland 1816, III, 231).

2 "Dans l'état actuel des choses, on voit s'agrandir, en gangnant sur la mer, les plaines humides" (Humboldt and Bonpland 1816, III, 232).

3 Lincoln Myers, conversation with the author, Gran Couva, Trinidad, July 2, 2011. Cf. Griffith and Oderson (2009, 21–86) and Leggett (2001, 24–27).

4 Leo Heileman, conversation with the author via Skype, July 19, 2011.

5 Heileman, conversation with the author, Gran Couva, January 4, 2013.

6 All quotations are from Angela Cropper, conversation with the author, Port of Spain, January 7, 2012.

7 Eden Shand, conversation with the author, Newark, Delaware, June 20, 2011.

8 Declaration of Barbados, Part One, Article III, Clause 2.

9 "Draft Protocol to the United Nations Framework Convention on Climate Change on Greenhouse Gas Emissions Reduction," submitted on May 17, 1996, as Paper No. 1 by Trinidad and Tobago on behalf of AOSIS for consideration by the Ad Hoc Group on the Berlin Mandate, fourth session, Geneva, July 9–16, 1996, http://unfccc.int/resource/docs/1996/agbm/misc02.pdf.

10 Surprisingly, in this period, the government invoked none of the available arguments, such as off-shoring, historical debt, or the distinction between subsistence and luxury emissions (cf. Agarwal and Narain 1992, 24ff.).

11 "Port of Spain Climate Change Consensus: The Commonwealth Climate Change Declaration," Port of Spain, November 28, 2009, Clause 13.

12 Cropper, remarks at the Commonwealth People's Forum, opening plenary session, Port of Spain, November 23, 2009.

13 Emily Gaynor Dick-Forde, remarks at the Heath, Safety, Security, and the Environment Conference, Port of Spain, September 29, 2009. The origins of the quotation are unclear.

14 Dick-Forde, remarks at the Commonwealth People's Forum, opening plenary session, Port of Spain, November 23, 2009.

15 Cropper, conversation with the author, Port of Spain, January 7, 2012.

16 Patrick Manning, conversation with the author, San Fernando, June 29, 2010.

17 Government of the Republic of Trinidad and Tobago, "Draft National Climate Change Policy for Trinidad and Tobago," 2010, 7.

18 The government held four meetings in total.

19 Shand, remarks at the National Consultation on Climate Change Policy, Port of Spain, March 23, 2010.

20 John Agard, remarks at the National Consultation on Climate Change Policy, Port of Spain, March 23, 2010.

21 Agard, conversation with the author, St. Augustine, Trinidad, January 29, 2010.

22 Agard was summarizing chapter 9 of a large report (UNEP 2007). He had been one of three lead coordinating authors of that chapter.

23 John Agard, "Environment in Development: From Plantation Economy, Biodiversity Loss and Global Warming towards Sustainable Development," lecture at the University of the West Indies, St. Augustine, February 25, 2010.

24 Remarks at Public Consultation on Climate Change Draft Policy, La Romaine, April 6, 2010.

25 Shyam Dyal, remarks at Public Consultation on Climate Change Draft Policy, La Romaine, April 6, 2010.

26 As presented by Garret Manwaring to the Health, Safety, Security and the Environment Conference of the American Chamber of Commerce of Trinidad and Tobago, Port of Spain, September 29, 2009.

27 Dyal, conversation with the author, Pointe-a-Pierre, March 3, 2010.

28 Remarks at Public Consultations on Climate Change Draft Policy, La Romaine, April 6, 2010.

29 Akilah Jaramogi, personal communication, Port of Spain, April 6, 2010.

30 Winston Rudder and Keisha Garcia, conversation with the author, Port of Spain, July 2, 2010.

31 Rudder, conversation with the author, Port of Spain, January 4, 2012.

32 Cropper, conversation with the author, Port of Spain, January 7, 2012.

Conclusion

1 She is quoting Svante Pääbo.

2 Remarks at the National Consultation on Climate Change Policy, Sangre Grande, April 20, 2010.

3 Johnny Stollmeyer, remarks to author, Port of Spain, January 5, 2010.

4 Dax Driver, remarks at the Green Business Forum, Port of Spain, March 24, 2011.

5 Joth Singh, remarks at the Green Business Forum, Port of Spain, March 24, 2011. Emphasis in original.

6 "Thousands Fill NYC Streets for Climate March," *Fox 5 News*, September 21, 2014, http://www.myfoxny.com/story/26588335/thousands-fill-nyc-streets -for-climate-march.

7 See Meadows (1998) for an earlier version of this argument.

8 All the energy statistics derive from Hermann (2006) and Hermann and Simon (2006). He uses the term *exergy* "as a common currency to assess and compare the reservoirs of theoretically extractable work we call energy resources" (Hermann 2006, 1685).

REFERENCES

Archivo General de Indias (AGI), Seville, Spain.

Audiencia de Caracas, legajos 150–53, 44, 444, 861.

Agard, J. B. R. 1988. "Petroleum Residues in Surficial Sediments from the Gulf of Paria, Trinidad." *Marine Pollution Bulletin* 19 (5): 231–33.

Agarwal, Anil, and Sunita Narain. 1992. *Towards a Green World.* New Delhi: Center for Science and the Environment.

Alatas, Syed Hussein. 1977. *The Myth of the Lazy Native: A Study of the Malays, Filipinos and Javanese from the 16th to the 20th Century and Its Function in the Ideology of Colonial Capitalism.* London: Frank Cass.

Andrews, Thomas G. 2008. *Killing for Coal: America's Deadliest Labor War.* Cambridge, MA: Harvard University Press.

Antoni, Robert. 2013. "A Counterfeit Utopia." *Cabinet* 51:60–68.

Appadurai, Arjun. 1986. *The Social Life of Things: Commodities in Cultural Perspective.* Cambridge: Cambridge University Press.

Appel, Hannah, Arthur Mason, and Michael Watts. 2015a. "Introduction: Oil Talk." In Appel, Mason, and Watts, 2015b, 1–26.

Appel, Hannah, Arthur Mason, and Michael Watts, eds. 2015b. *Subterranean Estates: Life Worlds of Oil and Gas.* Ithaca, NY: Cornell University Press.

Apter, Andrew. 2005. *The Pan-African Nation: Oil and the Spectacle of Culture in Nigeria.* Chicago: University of Chicago Press.

Armstrong, Franny. 2009. *The Age of Stupid* [film]. Spanner Films/Passion Pictures.

Auyero, Javier, and Débora Alejandra Swistun. 2009. *Flammable: Environmental Suffering in an Argentine Shantytown.* Oxford: Oxford University Press.

Bader, Christine. 2014. *The Evolution of a Corporate Idealist: When Girl Meets Oil.* Brookline, MA: Bibliomotion.

Baird, Juli. 2010. "Oil's Shame in Africa." *Newsweek* 156 (4): 16.

Barr, K. W., S. T. Waite, and C. C. Wilson. 1958. "The Mode of Oil Occurrence in the Miocene of Southern Trinidad, B.W.I." In *Habitat of Oil*, ed. Lewis G. Weeks, 533–50. Tulsa, OK: American Association of Petroleum Geologists.

Barrett, Ross, and Daniel Worden. 2014. "Introduction." In *Oil Culture*, ed. Ross Barrett and Daniel Worden, xvii–xxxiii. Minneapolis: University of Minnesota Press.

Barry, Andrew. 2006. "Technological Zones." *European Journal of Social Theory* 9:239–53.

Barry, Andrew. 2013. *Material Politics: Disputes along the Pipeline*. Chichester, UK: John Wiley.

Bass, Rick. 1989. *Oil Notes*. Boston: Houghton Mifflin.

Bass, Rick. 2012. *Oil Notes*, 2nd ed. Lincoln: University of Nebraska Press.

Beck, Ulrich. 1992. *Risk Society*. London: Sage.

Beecher, Jonathan. 1986. *Charles Fourier: The Visionary and His World*. Berkeley: University of California Press.

Beecher, Jonathan, and Richard Bienvenu, eds. 1971. *The Utopian Vision of Charles Fourier*. Boston: Beacon.

Benítez-Rojo, Antonio. 1992. *The Repeating Island: The Caribbean and the Postmodern Perspective*. Durham, NC: Duke University Press.

Bennett, Jane. 2010. *Vibrant Matter: A Political Ecology of Things*. Durham, NC: Duke University Press.

Besson, Gerard A. 2001. *The Angostura Historical Digest of Trinidad and Tobago*. Cascade, Trinidad and Tobago: Paria.

Besson, Gerard A. 2010. *The Cult of the Will*. Cascade, Trinidad and Tobago: Paria.

Bilby, Kenneth M. 2005. *True-Born Maroons*. Gainesville: University Press of Florida.

Bitterli, Peter. 1958. "Herrera Subsurface Structure of Penal Field, Trinidad, B.W.I." *Bulletin of the American Association of Petroleum Geologists* 42 (1): 145–58.

Black, Brian. 2000. *Petrolia: the Landscape of America's First Oil Boom*. Baltimore: Johns Hopkins University Press.

Bonnet, Natacha. 2008. "L'organisation du travail servile sur la sucrerie domingoise au XVIIIᵉ siècle." In *L'esclavage et les plantations: De l'établissement de la servitude à son abolition*, ed. Philippe Hrodēj. Rennes, France: Presses Universitaries de Rennes.

Borde, Pierre-Gustave-Louis. 1882. *Histoire de l'Île de la Trinidad sous le Gouvernement Espagnol*. Vol. 2. Paris: Maisonneuve.

Boym, Svetlana. 2008. *Architecture of the Off-Modern*. New York: Princeton Architectural Press.

Braun, Bruce. 2000. "Producing Vertical Territory: Geology and Governmentality in Late Victorian Canada." *Cultural Geographies* 7 (1): 9–46.

Breglia, Lisa. 2013. *Living with Oil: Promises, Peaks, and Declines on Mexico's Gulf Coast*. Austin: University of Texas Press.

Brereton, Bridget. 2001. *A History of Modern Trinidad, 1783–1962*. Oxford: Heinemann International.

Bridge, Gavin. 2004. "Gas, and How to Get It." *Geoforum* 35:395–97.

Bridge, Gavin, and Philippe Le Billon. 2013. *Oil*. Cambridge: Polity.

Bridge, Gavin, and Andew Wood. 2010. "Less Is More: Spectres of Scarcity and the Politics of Resource Access in the Upstream Oil Sector." *Geoforum* 41:565–76.

Brostowin, Patrick. 1969. "John Adolphus Etzler: Scientific-Utopian during the 1830's and 1840's." PhD dissertation, New York University.

Bullard, Robert D. 1990. *Dumping in Dixie: Race, Class, and Environmental Quality*. Boulder, CO: Westview.

Burtynsky, Edward. 2009. *Oil*. Göttingen, Germany: Steidl.

Butti, Ken, and John Perlin. 1980. *A Golden Thread: 2500 Years of Solar Architecture and Technology*. Palo Alto, CA: Cheshire.

Campbell, Jacob. 2014. "The Nature of Hydrocarbons: Industrial Ecology, Resource Depletion, and Politics of Renewability in Trinidad and Tobago." PhD diss., University of Arizona, Tucson.

Carlyle, Thomas. 1849. "Occasional Discourse on the Negro Question." *Fraser's Magazine for Town and Country* 40:670–79.

Carrera, Gustavo Luis. [1972] 2005. *La Novela del Petróleo en Venezuela*. Mérida, Venezuela: Universidad de los Andes.

Carrier, James G. 2010. "Protecting the Environment the Natural Way: Ethical Consumption and Commodity Fetishism." *Antipode* 42 (3): 672–89.

Carrington, Selwyn H. H. 2003. "Capitalism and Slavery and Caribbean Historiography: An Evaluation." *Journal of African American History* 88 (3): 304–12.

Carter, Paul. 1987. *The Road to Botany Bay: An Exploration of Landscape and History*. New York: Knopf.

Chakrabarty, Dipesh. 2009. "The Climate of History: Four Theses." *Critical Enquiry* 35:197–222.

Chakravarty, Shoibal, Ananth Chikkatur, Heleen de Coninck, Stephen Pacala, Robert Socolow, and Massimo Tavoni. 2010. "Sharing Global CO_2 Emissions Reductions among One Billion High Emitters." *Proceedings of the National Academy of Sciences* 106 (29): 11884–88.

Chamoiseau, Patrick. 1992. *Texaco*. Paris: Gallimard.

Chapman, Chelsea. 2013. "Multinational Resources: Ontologies of Energy and the Politics of Inevitability in Alaska." In *Cultures of Energy: Power, Practices, Technologies*, ed. Sarah Strauss, Stephanie Rupp, and Thomas Love, 96–109. Walnut Creek, CA: Left Coast.

Chase, Malcolm. 2011. "Exporting the Owenite Utopia: Thomas Powell and the Tropical Emigration Society." In *Robert Owen and His Legacy*, ed. Noel Thompson and Chris Williams, 198–217. Cardiff: University of Wales Press.

Claeys, Gregory. 1986. "John Adophus Etzler, Technological Utopianism, and British Socialism: The Tropical Emigration Society's Venezuelan Mission and Its Social Context, 1833–1848." *English Historical Review* 101 (399): 351–75.

Claeys, Gregory, and Lyman Tower Sargent. 1999. "Introduction." In *The Utopia Reader*, ed. Gregory Claeys and Lyman Tower Sargent. New York: New York University Press.

Cliff, Michelle. 1987. *No Telephone to Heaven*. New York: Dutton.

Climate One. 2011. "Blessed 350: Bill McKibben and Paul Hawken." September 8. http://envirobeat.com/?p=3604.

Cochrane, Alexander. 1805. "A Report on the Pitch Lake, etc., in the Island of Trinidad." Publication no. 301, Historical Society of Trinidad and Tobago.

Coll, Steven. 2012. *Private Empire: ExxonMobil and American Power*. New York: Penguin.

Collard, Rosemary-Claire. 2014. "Putting Animals Back Together, Taking Commodities Apart." *Annals of the Association of American Geographers* 104 (1): 151–65.

Colony of Trinidad. 1902. *Report of the Asphalt Industry Commission.* London: Waterlow and Sons.

Colony of Trinidad. 1903. *Report of Proceedings before Their Honors the Commissions Appointed to Enquire into Matters Concerning the Asphalt Industry at La Brea in the Island of Trinidad.* Port of Spain: Government Printing Office.

Coopersmith, Jennifer. 2010. *Energy, the Subtle Concept: The Discovery of Feynman's Blocks from Leibniz to Einstein.* Oxford: Oxford University Press.

Coronil, Fernando. 1997. *The Magical State: Nature, Money, and Modernity in Venezuela.* Chicago: University of Chicago Press.

Crate, Susan A. 2008. "Gone the Bull of Winter? Grappling with the Cultural Implications of and Anthropology's Role(s) in Global Climate Change." *Current Anthropology* 49 (4): 569–95.

Crone, G. R. 1938. "The Origin of the Name Antillea." *Geographical Journal* 91 (3): 260–62.

Cropper, Angela. 1994. "Small Is Vulnerable." *Our Planet* 6 (1): 9–12.

Cropper Foundation. 2008. "Mind Your Own Business: How to Keep Track of Trinidad and Tobago's Energy Billions." Port of Spain: Cropper Foundation.

Crosby, Alfred W. 2006. *Children of the Sun: A History of Humanity's Unappeasable Appetite for Energy.* New York: W. W. Norton.

Crutzen, Paul J. 2002. "Geology of Mankind." *Nature* 415 (January 3): 23.

Crutzen, P. J., and E. F. Stoermer. 2000. "The Anthropocene." *IGBP Newsletter* 41: 12–14.

Cudjoe, Selwyn. 2003. *Beyond Boundaries: The Intellectual Tradition of Trinidad and Tobago in the Nineteenth Century.* Wellesley, MA: Calaloux.

Cunningham-Craig, E. H. 1912. *Oil-Finding: An Introduction to the Geological Study of Petroleum.* London: Edward Arnold.

Curtin, Philip D. 1990. *The Rise and Fall of the Plantation Complex: Essays in Atlantic History.* New York: Cambridge University Press.

Davidov, Veronica. 2012. "Saving Nature or Performing Sovereignty: Ecuador's Initiative to 'Keep Oil in the Ground.'" *Anthropology Today* 28 (3): 12–15.

Davis, David Brion. 1966. *The Problem of Slavery in Western Culture.* Ithaca, NY: Cornell University Press.

Davis, David Brion. 2006. *Inhuman Bondage: The Rise and Fall of Slavery in the New World.* Oxford: Oxford University Press.

Davis, Mike. 2010. "Who Will Build the Ark?" *New Left Review* 61:29–61.

Debien, G. 1966. "Le marronage aux Antilles française au XVIIIe siècle." *Caribbean Studies* 6 (3): 3–43.

deGannes, Karen. 2013. "Environment, Development, and Citizenship: Narrative Processes as Environmental Revolution and Political Change in Post-colonial Trinidad and Tobago." PhD diss., University of Michigan.

Deloughrey, Elizabeth. 2004. "Island Ecologies and Caribbean Literatures." *Tijdschrift voor Economische en Sociale Geografie* 95 (3): 298–310.

de Sousa, Luis. 2008. "What Is a Human Being Worth (in Terms of Energy)?" Oil Drum: Europe, July 29. http://www.theoildrum.com/node/4315.

de Verteuil, Anthony. 1992. *Seven Slaves and Slavery: Trinidad, 1777–1838*. Port of Spain: Anthony de Verteuil.

de Verteuil, Anthony. 1994. *The Germans in Trinidad*. Port of Spain: Litho.

de Verteuil, Anthony. 1996. "Bunsee Partap (and an Account of the Dome Fire, 1928)." Appendix to *A History of Trinidad Oil*, by George E. Higgins, 394–410. Port of Spain: Trinidad Express Newspapers.

de Verteuil, Anthony. 2002. *Western Isles of Trinidad*. Port of Spain: Litho.

de Verteuil, L. A. A. 1848. "Essay on the Cultivation of the Sugar-Cane in Trinidad." In *Three Essays on the Cultivation of the Sugar-Cane in Trinidad*. Port of Spain: The Standard.

de Verteuil, L. A. A. 1858. *Trinidad: Its Geography, Natural Resources, Administration, Present Condition and Prospects*. London: Ward and Lock.

Driver, Thackwray. 1998. "The Theory and Politics of Mountain Rangeland Conservation and Pastoral Development in Colonial Lesotho." PhD diss., University of London.

Driver, Thackwray. 1999. "Anti-erosion Policies in the Mountain Areas of Lesotho: The South African Connection." *Environment and History* 5:1–25.

Driver, Thackwray. 2002. "Watershed Management, Private Property and Squatters in the Northern Range, Trinidad." *IDS Bulletin* 33 (1): 84–93.

Dukes, Jeffrey S. 2003. "Burning Buried Sunshine: Human Consumption of Ancient Solar Energy." *Climatic Change* 61: 31–44.

Dundonald, Thomas Cochrane. 1851. *Notes on the Mineralogy, Government, and Conditions of the British West India Islands and North-American Maritime Colonies*. London: James Ridgway.

Dunn, Richard S. 1972. *Sugar and Slaves: The Rise of the Planter Class in the English West Indies*. Chapel Hill: University of North Carolina Press.

Dunn, Richard S. 2014. *A Tale of Two Plantations: Slave Life and Labor in Jamaica and Virginia*. Cambridge, MA: Harvard University Press.

Energy Information Administration. 2014. *June 2014: Monthly Energy Review*. Washington, DC: U.S. Energy Information Administration.

Etzler, John. 1833. *The Paradise within the Reach of All Men, without Labour, by Powers of Nature and Machinery*. Part 1. Pittsburgh: Etzler and Reinhold.

Etzler, John. 1841. *The New World or Mechanical System, to Perform the Labours of Man and Beast by Inanimate Powers, That Cost Nothing, for Producing and Preparing the Substances of Life*. Philadelphia: C. F. Stollmeyer.

Etzler, John. 1844a. "Emigration to the Tropical World for the Melioration of All Classes of People of All Nations." Surrey, UK: Concordium.

Etzler, John. 1844b. "Two Visions of J. A. Etzler: A Revelation of Futurity." Surrey, UK: Concordium.

Ewalt, Margaret R. 2008. *Peripheral Wonders: Nature, Knowledge, and Enlightenment in the Eighteenth-Century Orinoco*. Lewisburg, PA: Bucknell University Press.

Fassin, Didier. 2009. *The Empire of Trauma: An Enquiry into the Condition of Victimhood.* Translated by Rachel Gomme. Princeton, NJ: Princeton University Press.

Ferber, Edna. 1952. *Giant.* Garden City, NY: Doubleday.

Ferguson, James. 1990. *The Anti-Politics Machine: "Development," Depoliticization, and Bureaucratic Power in Lesotho.* Cambridge: Cambridge University Press.

Ferguson, James. 2005. "Seeing Like an Oil Company: Space, Security, and Global Capital in Neoliberal Africa." *American Anthropologist* 107 (3): 377–82.

Finer, Matt, Remi Moncel, and Clinton N. Jenkins. 2010. "Leaving the Oil under the Amazon: Ecuador's Yasuní-ITT Initiative." *Biotropica* 42 (1): 63–66.

Fourier, Charles. 1840. *Le nouveau monde industriel et sociétaire, ou les séries passionées.* Vol. 1. Bruxelles.

Fraser, Lionel Mordaunt. 1971. *History of Trinidad.* Vol. 2. London: Frank Cass.

Garrard, Greg. 2004. *Ecocriticism.* Abington, UK: Routledge.

Gelber, Elizabeth. 2015. "Black Oil Business: Rogue Pipelines, Hydrocarbon Dealers, and the 'Economics' of Oil Theft." In Appel, Mason, and Watts, 2015b, 274–90.

Gerbi, Antonello. [1955] 1973. *The Dispute of the New World.* Translated by Jeremy Moyle. Pittsburgh: University of Pittsburgh Press.

Ghosh, Amitav. 1992. "Petrofiction: The Oil Encounter and the Novel." *New Republic,* March 2, 29–34.

Gillis, John R. 2004. *Islands of the Mind.* New York: Palgrave Macmillan.

Glacken, Clarence J. 1967. *Traces on the Rhodian Shore: Nature and Culture in Western Thought from Ancient Times to the End of the Eighteenth Century.* Berkeley: University of California Press.

Goldstein, Daniel M. 2005. "Flexible Justice: Neoliberal Violence and 'Self-Help' Security in Bolivia." *Critique of Anthropology* 25 (4): 389–411.

Goldstein, Daniel. 2012. *Outlawed: Between Security and Rights in a Bolivian Shantytown.* Durham, NC: Duke University Press.

Gómez, Nicolás Wey. 2008. *The Tropics of Empire: Why Columbus Sailed South.* Cambridge, MA: MIT Press.

Gore, Al. 2006. *An Inconvenient Truth: A Global Warning* [film]. Hollywood, CA: Paramount Pictures.

Graves, John. 1995. "Introduction" to *Oil Notes,* by Rick Bass, 2nd ed. Dallas: Southern Methodist University Press.

Griffith, Mark D., and Derrick Oderson. 2009. *Nuts and Bolts.* St. Michael, Barbados: CaribInvest.

Grove, Richard H. 1995. *Green Imperialism: Colonial Expansion, Tropical Island Edens and the Origins of Environmentalism, 1600–1860.* Cambridge: Cambridge University Press.

Gumilla, Joseph. [1745] 1945. *El Orinoco Ilustrado.* Edited by Constantino Bayle. Madrid: M. Aguilar.

Gumilla, Joseph. 1970. *P. José Gumilla: Escritos Varios.* Edited by José del Rey. Caracas: Academia Nacional de la Historia.

Hager, Thomas. 2008. *The Alchemy of Air*. New York: Harmony.

Hall, Francis. 1827. *Colombia: Its Present State in Respect of Climate, Soil, Productions, Population, Government, Commerce, Revenue, Manufactures, Arts, Literature, Manners, Education, and Inducements to Emigration*. London: Baldwin, Cradock, and Joy.

Hamilton, Clive. 2013. *Earthmasters*. New Haven, CT: Yale University Press.

Hayes, Christopher. 2014. "The New Abolitionism." *The Nation*, April 29.

Heileman, Leo. 1993. "The Alliance of Small Island States (AOSIS): A Mechanism for Coordinated Representation of Small Island States on Issues of Common Concern." *Ambio* 22 (1): 55–56.

Hein, Carola. 2009. "Global Landscapes of Oil." *New Geographies* 2:33–42.

Heise, Ursula K. 2008. *Sense of Place and Sense of Planet*. Oxford: Oxford University Press.

Heringman, Noah. 2004. *Romantic Rocks, Aesthetic Geology*. Ithaca, NY: Cornell University Press.

Hermann, Weston A. 2006. "Quantifying Exergy Resources." *Energy* 31:1685–1702.

Hermann, Weston A., and A. J. Simon. 2006. "Global Exergy Flux, Reservoirs, and Destruction." Stanford, CA: Global Climate and Energy Project, Stanford University. http://www.gcep.stanford.edu/pdfs/GCEP_Exergy_Poster_web.pdf.

Higgins, George E. 1996. *A History of Trinidad Oil*. Port of Spain: Trinidad Express Newspapers.

Higman, B. W. 2000. "The Sugar Revolution." *Economic History Review* 53 (2): 213–36.

Hirschman, Albert. 1971. *A Bias for Hope: Essays on Development and Latin America*. New Haven, CT: Yale University Press.

Hitchcock, Peter. 2010. "Oil in an American Imaginary." *New Formations* 69:81–97.

Hosein, Gabrielle. 2007. "Survival Stories: Challenges Facing Youth in Trinidad and Tobago." *Race and Class* 49 (2): 125–30.

Hubbert, M. King. 1962. "Energy Resources: A Report to the Committee on Natural Resources." Washington, DC: National Academy of Sciences, National Research Council.

Huber, Matthew. 2012. "Refined Politics: Petroleum Products, Neoliberalism, and the Ecology of Entrepreneurial Life." *Journal of American Studies* 46 (2): 295–312.

Huber, Matthew. 2013. *Lifeblood: Oil, Freedom, and the Forces of Capital*. Minneapolis: University of Minnesota Press.

Hughes, David McDermott. 2010. *Whiteness in Zimbabwe: Race, Landscape, and the Problem of Belonging*. New York: Palgrave Macmillan.

Humboldt, Alexandre de, and Aimé Bonpland. 1805. *Essai sur la géographie des plantes*. Paris: Levraut, Schoell, et Compagnie.

Humboldt, Alexandre de, and Aimé Bonpland. 1816. *Voyage aux Régions Équinoxiales du Nouveau Continent Fait en 1799, 1800, 1801, 1802, 1803, et 1804*. Vol. 3. Paris: Librairie Grecque.

Hyne, Norman J. 1995. *Nontechnical Guide to Petroleum Geology, Exploration, Drilling, and Production*. Tulsa, OK: PennWell.

Illich, Ivan. [1983] 2009. "The Social Construction of Energy." *New Geographies* 2:11–23.

Institute of Marine Affairs. 2003. "Environmental Impact Assessment for the Establishment of an Industrial Estate at Union Estate, La Brea (Phase 2), Southwestern Trinidad." Chaguaramas, Trinidad and Tobago: Institute of Marine Affairs.

International Energy Agency. 2010. *CO_2 Emissions from Fuel Combustion: Highlights.* Paris: OECD/IEA.

Jacobson, Mark Z., and Mark A. Delucchi. 2009. "A Path to Sustainable Energy by 2030." *Scientific American* 301 (5): 58–65.

James, C. L. R. 1938. *The Black Jacobins.* London: Secker and Warburg.

James, C. L. R. [1938] 1989. *The Black Jacobins: Toussaint L'Ouverture and the San Domingo Revolution.* New York: Vintage.

James, C. L. R. 1963a. *Beyond a Boundary.* New York: Pantheon.

James, C. L. R. 1963b. *The Black Jacobins,* 2nd ed. New York: Vintage.

James, C. L. R. 1966. "Rohan Kanhai: A Study in Confidence." *New World Quarterly* 3 (1): 13–15.

John, A. Meredith. 1988. *The Plantation Slaves of Trinidad, 1783–1816: A Mathematical and Demographic Enquiry.* Cambridge: Cambridge University Press.

Jørgensen, Dolly. 2014. "Mixing Oil and Water: Naturalizing Offshore Oil Platforms in American Aquariums." In *Oil Culture,* ed. Ross Barrett and Daniel Worden, 267–88. Minneapolis: University of Minnesota Press.

Joseph, E. L. 1838. *History of Trinidad.* London: A. K. Newman and Co.

Kapuściński, Ryszard. 1986. *Shah of Shahs.* New York: Vintage.

Kashi, Ed, and Michael Watts. 2008. *Curse of the Black Gold: 50 Years of Oil in the Niger Delta.* Brooklyn, NY: PowerHouse.

Kelman, Ilan. 2010. "Hearing Local Voices from Small Island Developing States for Climate Change." *Local Environment* 15 (7): 605–19.

Kempadoo, Oonya. 2001. *Tide Running.* Boston: Beacon.

Kenny, Julian. 2011. *Of Dragons and Doves: Essays of Our Times.* Arima, Trinidad and Tobago: University of Trinidad and Tobago.

Khan, Aisha. 1997. "Rurality and 'Racial' Landscapes in Trinidad." In *Knowing Your Place: Rural Identity and Cultural Hierarchy,* ed. Barbara Chinge and Gerald W. Creed, 39–69. New York: Routledge.

Khan, Aisha. 2001. "Journey to the Center of the Earth: The Caribbean as Master Symbol." *Cultural Anthropology* 16 (3): 271–302.

Kidron, Carol A. 2009. "Toward an Ethnography of Silence: The Lived Presence of the Past in the Everyday Life of Holocaust Trauma Survivors and Their Descendants in Israel." *Current Anthropology* 50 (1): 5–19.

Klein, Naomi. 2014. *This Changes Everything: Capitalism vs. the Climate.* New York: Simon and Schuster.

Kolbert, Elizabeth. 2014. *The Sixth Extinction: An Unnatural History.* New York: Holt.

Konrad, John, and Tom Shroder. 2011. *Fire on the Horizon: The Untold Story of the Gulf Oil Disaster.* New York: HarperCollins.

Kormann, Carolyn. 2013. "Scenes from a Melting Planet: On the Climate-Change Novel." *New Yorker*, July 3.

Korngold, Ralph. 1944. *Citizen Toussaint*. Boston: Little, Brown.

Kövecses, Zoltán. 2002. *Metaphor: A Practical Introduction*. Oxford: Oxford University Press.

Kropotkin, P. N. 1997. "On the History of Science Professor N. A. Koudryavtsev (1893–1971) and the Development of the Theory of the Origin of Oil and Gas." *Earth Sciences History* 16 (1): 17–20.

Kublalsingh, Wayne. 2009. *Ital Revolution*. Toronto: Just World.

Kublalsingh, Wayne. 2011. "An Ecological Messiah." In *Of Dragons and Doves: Essays of Our Times*, by Julian Kenny, 308–9. Arima, Trinidad and Tobago: University of Trinidad and Tobago.

Kublalsingh, Wayne. 2014. "Global Village or Global Empire: Twenty Short Essays." Unpublished manuscript.

Kuhn, Thomas S. 1962. *The Structure of Scientific Revolutions*. Chicago: University of Chicago Press.

Kunstler, James Howard. 2005. *The Long Emergency: Surviving the Converging Catastrophes of the Twenty-First Century*. New York: Atlantic Monthly.

Labban, Mazen. 2010. "Oil in Parallax: Scarcity, Markets, and the Financialization of Accumulation." *Geoforum* 41:541–52.

Lafargue, Paul. [1880] 1994. *Le droit à la paresse*. Paris: Éditions Mille et Une Nuit.

Larkin, Brian. 2013. "The Politics and Poetics of Infrastructure." *Annual Review of Anthropology* 42:327–43.

Latour, Bruno. 1999. *Pandora's Hope: Essays on the Reality of Science Studies*. Cambridge, MA: Harvard University Press.

Latour, Bruno. 2005. *Reassembling the Social: An Introduction to Actor-Network Theory*. Oxford: Oxford University Press.

Lazrus, Heather. 2009. "The Governance of Vulnerability: Climate Change and Agency in Tuvalu, South Pacific." In *Anthropology and Climate Change: From Encounters to Actions*, ed. Susan A. Crate and Mark Nuttall, 240–49. Walnut Creek, CA: Left Coast.

Lazrus, Heather. 2012. "Sea Change: Island Communities and Climate Change." *Annual Review of Anthropology* 41:285–301.

Leggett, Jeremy. 2001. *The Carbon War*. New York: Routledge.

LeMenager, Stephanie. 2012. "Fossil, Fuel: Manifesto for the Post-oil Museum." *Journal of American Studies* 46 (2): 375–94.

LeMenager, Stephanie. 2014. *Living Oil: Petroleum Culture in the American Century*. Oxford: Oxford University Press.

Lewis, Martin W., and Kären E. Wigen. 1997. *The Myth of Continents*. Berkeley: University of California Press.

Li, Tania Murray. 2000. "Articulating Indigenous Identity in Indonesia: Resource Politics and the Tribal Slot." *Comparative Studies in Society and History* 42 (1): 149–79.

Li, Tania Murray. 2007. *The Will to Improve: Governmentality, Development, and the Practice of Politics*. Durham, NC: Duke University Press.

Liverpool, Hollis "Chalkdust." 2001. *Rituals of Power and Rebellion: The Carnival Tradition in Trinidad and Tobago, 1763–1962*. Chicago: Research Associates School Times.

Lloyd, Chistopher. 1947. *Lord Cochrane: Seaman, Radical, Liberator*. New York: Henry Holt.

Logan, Joshua. n.d. *The Price of Progress*. Unpublished play.

Lovelace, Earl. 1979. *The Dragon Can't Dance*. London: Andre Deutsch.

Lovelace, Earl. 1984. *Jestina's Calypso and Other Plays*. London: Heinemann.

Lyell, Charles. 1830. *Principles of Geology*. Vol. 1. London: John Murray.

Maisier, Véronique. 2015. *Violence in Caribbean Literature: Stories of Stones and Blood*. Lanham, MD: Lexington.

Mallet, F. 1802. *Descriptive Account of the Island of Trinidad*. London.

Malm, Andreas, and Alf Hornborg. 2014. "The Geology of Mankind? A Critique of the Anthropocene Narrative." *Anthropocene Review* 1 (1): 62–69.

Mankekar, Purnima. 2004. "Dangerous Desires: Television and Erotics in Late Twentieth-Century India." *Journal of Asian Studies* 63 (2): 403–31.

Mann, Charles C. 2013. "What If We Never Run Out of Oil?" *Atlantic*, April 24.

Manyena, Siambabala Bernard, Geoff O'Brien, Phil O'Keefe, and Joanne Rose. 2011. "Disaster Resilience: A Bounce Back or Bounce Forward Ability?" *Local Environment* 16 (5): 417–24.

Marcone, Jorge. 2013. "Humboldt in the Orinoco and the Environmental Humanities." *Hispanic Issues on Line* 12:76–91.

Martin, Gaston. 1948. *Histoire de l'Esclavage dans les Colonies Françaises*. Paris: Presses Universitaires de France.

Martínez, A. R., D. C. Jon, H. Dekker, and Shofner Smith. 1987. "Classification and Nomenclature Systems for Petroleum and Petroleum Reserves: 1987 Report." World Petroleum Congress.

Marx, Karl. 1976. *Capital: A Critique of Political Economy*. Vol. 1. Translated by Ben Fowkes. New York: Vintage.

Marx, Leo. 1964. *The Machine in the Garden: Technology and the Pastoral Ideal in America*. Oxford: Oxford University Press.

Massé, Armand. 1988. *The Diaries of Abbé Armand Massé, 1878–1883*. Vol. 3. Translated by M. L. de Verteuil. Port of Spain: M. L. de Verteuil.

Mathews, Andrew. 2014. "Domesticating the Global Carbon Cycle through Italian Forests." Paper presented at the annual meeting of the American Anthropological Association, Washington, DC, December 3–7.

Mathieson, William Law. 1926. *British Slavery and Its Abolition, 1823–1838*. London: Longmans.

Maurer, Bill. 2006. "The Anthropology of Money." *Annual Review of Anthropology* 35:15–36.

McGlade, Christophe, and Paul Elkins. 2015. "The Geographical Distribution of Fossil Fuels Unused When Limiting Global Warming to 2˚C." *Nature* 517:187–93.

McKelvey, V. E. 1972. "Mineral Resource Estimates and Public Policy." *American Scientist* 60:32–40.

McKibben, Bill. 1989. *The End of Nature*. New York: Anchor.

McKibben, Bill. 2003. "Worried? Us?" *Granta* 83:8–12.

McKibben, Bill. 2010. *Eaarth: Making a Life on a Tough New Planet*. New York: Times Books.

McKibben, Bill. 2012. "Global Warming's Terrifying New Math." *Rolling Stone*, August 2.

McNeill, J. R. 2000. *Something New under the Sun: An Environmental History of the Twentieth-Century World*. New York: W. W. Norton.

McPhee, John. 1980. *Basin and Range*. New York: Farrar, Straus and Giroux.

Meadows, Donella. 1998. "Thomas Jefferson and Donella Meadows, Slave-Owners." Donella Meadows Institute, November 12. http://www.donellameadows .org/archives/thomas-jefferson-and-donella-meadows-slave-owners/.

Menard, Russell R. 2006. *Sweet Negotiations: Sugar, Slavery, and Plantation Agriculture in Early Barbados*. Charlottesville: University of Virginia Press.

Mendes, Alfred. 1934. *Pitch Lake*. London: Duckworth.

Merry, Sally Engel. 2011. "Measuring the World: Indicators, Human Rights, and Global Governance." *Current Anthropology* 52 (supplement 3): 83–95.

Messerly, Oscar. 1902. *Some Contributions to the Scientific Study of the Asphaltic Deposits of the Island of Trinidad and Neighbouring Mainland with General Considerations on the Judicial Questions Connected with the Exploitation of Asphalt in Trinidad*. Trinidad: Oscar Messerly.

Metz, Bert, Ogunlade Loos, and Leo Meyers, eds. 2005. *Carbon Dioxide Capture and Storage*. Cambridge: Cambridge University Press.

Miller, Daniel. 1994. *Modernity, an Ethnographic Approach: Dualism and Mass Consumption in Trinidad*. Oxford: Berg.

Miller, Daniel. 2011. *Tales from Facebook*. London: Polity.

Ministry of Energy and Energy Industries. 2009. *The Republic of Trinidad and Tobago: Celebrating a Century of Commercial Oil Production*. London: FIRST.

Mintz, Sidney W. 1985. *Sweetness and Power: The Place of Sugar in Modern History*. New York: Penguin.

Mitchell, Timothy. 2009. "Carbon Democracy." *Economy and Society* 38 (3): 399–432.

Mitchell, Timothy. 2011. *Carbon Democracy: Political Power in the Age of Oil*. New York: Verso.

Mollat, Michel. 1965. "Soleil et navigation au temps des découvertes." In *Le Soleil à la Renaissance: Sciences et Mythes*. Brussels: Presses Universitaires de Bruxelles.

Moore, Amelia. 2010. "Climate Changing Small Islands: Considering Social Science and the Production of Island Vulnerability and Opportunity." *Environment and Society: Advances in Research* 1:116–31.

Moore, Michael, dir. 2004. *Fahrenheit 9/11* [film]. Lionsgate Films.

Moors, Kent. 2011. *The Vega Factor: Oil Volatility and the Next Global Crisis.* Hoboken, NJ: Wiley.

Mottley, Wendell. 2008. *Trinidad and Tobago: Industrial Policy, 1959–2008.* Kingston, Jamaica: Ian Randle.

Mouhot, Jean-François. 2011. "Past Connections and Present Similarities in Slave Ownership and Fossil Fuel Usage." *Climate Change* 105:329–55.

Mulchansingh, Vernon C. 1971. "The Oil Industry in the Economy of Trinidad." *Caribbean Studies* 11 (1): 73–100.

Mumford, Lewis. 1934. *Technics and Civilization.* New York: Harcourt, Brace.

Munif, Abdelrahman. 1994. *Cities of Salt.* Translated by Peter Theroux. New York: Vintage.

Nader, Laura. 1974. "Up the Anthropologist—Perspectives Gained from Studying Up." In *Reinventing Anthropology*, ed. Dell Hymes, 284–311. New York: Vintage.

Nader, Laura. 2004. "The Harder Path—Shifting Gears." *Anthropological Quarterly* 77 (4): 771–91.

Naipaul, V. S. 1962. *The Middle Passage.* New York: Vintage.

Naipaul, V. S. 1969. *The Loss of El Dorado.* New York: Knopf.

Naipaul, V. S. 1970. "Power to the Caribbean People." *New York Review of Books* 15 (4): 32–34.

Naipaul, V. S. 1988. *The Enigma of Arrival.* New York: Vintage.

Naipaul, V. S. 1994. *A Way in the World.* London: Heinemann.

Newell, Jennifer. 2010. *Trading Nature: Tahitians, Europeans, and Ecological Exchange.* Honolulu: University of Hawai'i Press.

Newson, Linda. 1976. *Aboriginal and Spanish Colonial Trinidad: A Study in Culture Contact.* London: Academic Press.

Nikiforuk, Andrew. 2012. *The Energy of Slaves: Oil and the New Servitude.* Vancouver: Greystone.

Nixon, Rob. 2011. *Slow Violence and the Environmentalism of the Poor.* Cambridge, MA: Harvard University Press.

Noel, Jesse. 1972. *Trinidad, Provincia de Venezuela: Historia de la Adminstración Española de Trinidad.* Caracas: Academia Nacional de Historia.

Norgaard, Kari. 2006. "'We Don't Really Want to Know': Environmental Justice and Socially Organized Denial of Global Warming in Norway." *Organization and Environment* 19 (3): 347–70.

Nugent, Dr. 1811. "Account of the Pitch Lake of the Island of Trinidad." *Transactions of the Geological Society of London* 1:63–76.

Nydahl, Joel. 1977. "Introduction." In *The Collected Works of John Adolphus Etzler*, by John Adolphus Etzler, vii–xxxi. Delmar, NY: Scholars' Facsimiles and Reprints.

Nye, David E. 1994. *American Technological Sublime.* Cambridge, MA: MIT Press.

Olien, Roger M., and Diana Davids Olien. 2000. *Oil and Ideology: The Cultural Creation of the American Petroleum Industry.* Chapel Hill: University of North Carolina Press.

Orwell, George. 1937. *The Road to Wigan Pier*. London: Victor Gollancz.

Pantin, Dennis. 2008. "A Sustainable Development Planning Framework for Mega-projects in Small Places." Unpublished ms. St. Augustine, Trinidad and Tobago.

Patterson, Orlando. 1967. *The Sociology of Slavery*. London: MacGibbon and Kee.

Pérez de Tudela, Juan, Carlos Seco Serrano, Ramón Ezquerra Abadía, and Emilio López Oto, eds. 1994. *Colección Documental del Descubrimiento (1470–1506)*. Vol. 3. Madrid: Editorial MAPFRE.

Perrow, Charles. 1999. *Normal Accidents: Living with High-Risk Technologies*. Princeton, NJ: Princeton University Press.

Persad, Krishna M. 2011. *The Petroleum Geology and Geochemistry of Trinidad and Tobago*. Southern Energy Research Centre.

Persad, Krishna M., and Mia M. Persad. 1993. *The Petroleum Encyclopedia of Trinidad and Tobago*. Krishna Persad.

Polanyi, Karl. 1944. *The Great Transformation: The Political and Economic Origins of Our Time*. Boston: Beacon.

Pratt, Mary Louise. 1992. *Imperial Eyes: Travel Writing and Transculturation*. London: Routledge.

Price, Richard. 1979. "Introduction: Maroons and Their Communities." In *Maroon Societies: Rebel Slave Communities in the Americas*, ed. Richard Price, 1–30. Baltimore: Johns Hopkins University Press.

Priest, Tyler. 2007. *The Offshore Imperative: Shell Oil's Search for Petroleum in Postwar America*. College Station: Texas A&M Press.

Quammen, David. 1996. *The Song of the Dodo: Island Biogeography in an Age of Extinctions*. New York: Scribner.

Rabinow, Paul. 1977. *Reflections on Fieldwork in Morocco*. Berkeley: University of California Press.

Ramos Pérez, Demetrio. 1958. "Un plan de inmigración y libre comercio defendido por Gumilla para Guayana en 1739." *Anuario de Estudios Americanos* 15:201–24.

Rapid Environmental Assessments. 2006. "Environmental Impact Assessment for the Proposed Establishment of an Aluminium Complex at Main Site North, Union Industrial Estate, Trinidad." Port of Spain: Rapid Environmental Assessments.

Renwick, David. 2008a. "The Krishna Persad Vision Machine Grinds On." *Energy Caribbean* 38:20.

Renwick, David. 2008b. "Natural Gas Audit: Reserves Being Replaced on a Yearly Basis." *Energy Caribbean* 39:16.

Renwick, David. 2009. "The Ryder Scott Reserves Audit: What Does It Tell Us?" *Contact* (Trinidad and Tobago Chamber of Commerce and Industry) 9 (4): 74.

Ribot, Jesse C. 1995. "The Causal Structure of Vulnerability: Its Application to Climate Impact Analysis." *GeoJournal* 35 (2): 119–22.

Ribot, Jesse C. 2009. "Vulnerability Does Not Just Fall from the Sky: Toward a Multi-scale Pro-poor Climate Policy." In *Social Dimensions of Climate Change: Equity and Vulnerability in a Warming World*, ed. Robin Mearns and Andrew Norton, 47–74. Washington, DC: World Bank.

Righter, Robert W. 2002. "Exoskeletal Outer Space Creations." In *Wind Power in View: Energy Landscapes in a Crowded World,* ed. Martin J. Pasqualetti, Paul Gipe, and Robert W. Righter, 19–41. San Diego: Academic Press.

Rival, Laura. 2010. "Ecuador's Yasuní-ITT Initiative: The Old and New Values of Petroleum." *Ecological Economics* 70:358–65.

Robert, J. Timmons, and Bradley C. Parks. 2007. *A Climate of Injustice: Global Inequality, North-South Politics, and Climate Policy.* Cambridge, MA: MIT Press.

Roberts, Justin. 2006. "Working between the Lines: Labor and Agriculture on Two Barbadian Sugar Plantations, 1796–97." *William and Mary Quarterly,* 3rd ser., 63 (3): 551–86.

Roberts, Paul. 2004. *The End of Oil: On the Edge of a Perilous New World.* Boston: Houghton Mifflin.

Rohlehr, Gordon. 1992. *My Strangled City and Other Essays.* San Juan, Trinidad and Tobago: Longman Trinidad.

Ruddiman, William F. 2013. "The Anthropocene." *Annual Review of Earth and Planetary Sciences* 41:45–68.

Rudiak-Gould, Peter. 2011. "Climate Change Mitigation and Self-Blame in the Marshall Islands." Paper presented at the annual meeting of the American Anthropological Association, Montreal, November 16–20.

Rudwick, Martin J. S. 1976. "The Emergence of a Visual Language for Geological Science, 1760–1840." *History of Science* 14:149–95.

Rudwick, Martin J. S. 2008. *Worlds before Adam: The Reconstruction of Geohistory in the Age of Reform.* Chicago: University of Chicago Press.

Ruiz, Juan Martínez. 2002. *El Lenguaje del Suelo (Toponimia).* Jaén, Spain: Universidade de Jaén.

Salgado, Sebastião. 1993. *Workers: An Archaeology of the Industrial Age.* New York: Aperture.

Sankeralli, Burton. 2009. *The RAG File: Writings of the Aluminium Smelter Wars.* Toronto: Just World.

Santayana, George. [1922] 1968. "Marginal Notes on Civilization in the United States." In *Santayana on America,* 188–92. New York: Harcourt, Brace and World.

Sawyer, Suzana. 2004. *Crude Chronicles: Indigenous Politics, Multinational Oil, and Neoliberalism in Ecuador.* Durham, NC: Duke University Press.

Sawyer, Suzana. 2010. "Human Energy." *Dialectical Anthropology* 34:67–75.

Schama, Simon. 1995. *Landscape and Memory.* New York: Knopf.

Scheper-Hughes, Nancy. 1995. "The Primacy of the Ethical: Propositions for a Militant Anthropology." *Current Anthropology* 36 (3): 409–20.

Scott, Heidi V. 2008. "Colonialism, Landscape, and the Subterranean." *Geography Compass* 2 (6): 1853–69.

Scott, James C. 1998. *Seeing Like a State: How Certain Schemes to Improve the Human Condition Have Failed.* New Haven, CT: Yale University Press.

Scott, Lawrence. 2012. *Light Falling on Bamboo*. Birmingham, UK: Tindal Street.

Sharife, Khadija. 2011. "Colonizing Africa's Atmospheric Commons." *Capitalism Nature Socialism* 22 (4): 74–92.

Shaxson, Nicholas. 2007. *Poisoned Wells: The Dirty Politics of African Oil*. New York: Palgrave Macmillan.

Sheller, Mimi. 2014. *Aluminum Dreams: The Making of Light Modernity*. Cambridge, MA: MIT Press.

Sheridan, Richard B. 1972. "Africa and the Caribbean in the Atlantic Slave Trade." *American Historical Review* 77 (1): 15–35.

Shiva, Vandana. 1992. "The Greening of Global Reach." *Ecologist* 22 (6): 258–59.

Simmons, Matthew R. 2005. *Twilight in the Desert: The Coming Saudi Oil Shock and the World Economy*. Hoboken, NJ: Wiley.

Sinclair, Upton. 1917. *King Coal: A Novel*. New York: Macmillan.

Sinclair, Upton. 1926. *Oil!* New York: Albert and Charles Boni.

"Sinking without a Trace: Australia's Climate Change Victims." 2008. *The Independent* (London), May 5.

Smil, Vaclav. 1994. *Energy in World History*. Boulder, CO: Westview.

Smil, Vaclav. 2008. *Energy in Nature and Society: General Energetics of Complex Systems*. Cambridge, MA: MIT Press.

Smil, Vaclav. 2010. *Energy Transitions: History, Requirements, Prospects*. Santa Barbara, CA: Praeger.

Society of Petroleum Engineers. 2001. "Guidelines for the Evaluation of Petroleum Reserves and Resources." Richardson, TX: Society of Petroleum Engineers.

Society of Petroleum Engineers, American Association of Petroleum Geologists, World Petroleum Council, Society of Petroleum Evaluation Engineers, and Society of Exploration Geophysicists. 2011. "Guidelines for Application of the Petroleum Resources Management System."

Soler, Rosario Sevilla. 1988. *Inmigración y Cambio Socio-Economico en Trinidad (1783–1797)*. Seville, Spain: Escuela de Estudios Hispano-Americanos.

Stafford, Robert A. 1990. "Annexing the Landscapes of the Past: British Imperial Geology in the Nineteenth Century." In *Imperialism and the Natural World*, ed. John MacKenzie, 67–89. Manchester, UK: Manchester University Press.

Stoll, Steven. 2008. *The Great Delusion: A Mad Inventor, Death in the Tropics, and the Utopian Origins of Economic Growth*. New York: Hill and Wang.

Stollmeyer, Conrad Friedrich. 1839. "Forrede." In *Der Himmel auf Erden oder Weg zur Glückseligkeit*, by Christian Gotthilf Salzmann, v–xii. Philadephia: C. F. Stollmeyer.

Stollmeyer, Conrad Friedrich. 1845. "The Sugar Question Made Easy." London: Effingham Wilson.

Stover, Leon, ed. 1996. *The Time Machine: An Invention—a Critical Text of the 1895 London First Edition, with an Introduction and Appendices*. Jefferson, NC: McFarland.

"The Sugar Question." 1845. London: Smith, Elder, and Co.

Szeman, Imre. 2014. "Conclusion: On Energopolitics." *Anthropological Quarterly* 87 (2): 253–64.

Szeman, Imre, and Maria Whiteman. 2012. "Oil Imag(e)naries: Critical Realism and the Oil Sands." *Imaginations* 3 (2): 46–66.

Thompson, Alvin O. 2006. *Flight to Freedom: African Runaways and Maroons in the Americas.* Kingston, Jamaica: University of the West Indies Press.

Thongchai Winichakul. 1994. *Siam Mapped: A History of the Geo-Body of a Nation.* Honolulu: University of Hawai'i Press.

Thorsheim, Peter. 2006. *Inventing Pollution: Coal, Smoke, and Culture in Britain since 1800.* Athens: Ohio University Press.

Timsar, Rebecca Golden. 2015. "Oil, Masculinity, and Violence: Egbesu Worship in the Niger Delta of Nigeria." In Appel, Mason, and Watts, 2015b, 72–88.

Tourism Development Company. n.d. (ca. 2010). "The La Brea Pitch Lake." Port of Spain: Tourism Development Corporation.

Trinitrain. 2010. "Rapid Rail Project Information Fact Sheet." Port of Spain: Trinitrain.

Trouillot, Michel-Rolph. 1991. "Anthropology and the Savage Slot: The Poetics and Politics of Otherness." In *Recapturing Anthropology*, ed. Richard G. Fox, 17–44. Santa Fe, NM: School of American Research Press.

Tsing, Anna. 2012. "Contaminated Diversity in 'Slow Disturbance': Potential Collaborators for a Liveable Earth." In *Why Do We Value Diversity? Biocultural Diversity in Global Context*, ed. Gary Martin, Diana Mincyte, and Ursula Münster, 95–97. Munich: Rachel Carson Center for Environment and Society.

Tufte, Edward. 2006. *Beautiful Evidence.* Cheshire, CT: Graphics Press.

UNEP (United Nations Environment Programme). 2007. *Global Environmental Outlook (GEO) 4: Environment and Development.* New York: United Nations Environment Programme.

United Nations Statistics Division. 2009. *Energy Statistics Yearbook.* New York: United Nations Statistics Division.

Wahab, Amar. 2010. *Colonial Inventions: Landscape, Power, and Representation in Nineteenth-Century Trinidad.* Newcastle upon Tyne, UK: Cambridge Scholars.

Wainwright, Joel, and Geoff Mann. 2013. "Climate Leviathan." *Antipode* 45 (1): 1–22.

Walcott, Derek. 1992. *The Antilles: Fragments of an Epic Memory.* New York: Farrar, Straus and Giroux.

Walcott-Hackshaw, Elizabeth. 2007. *Four Taxis Facing North.* Hexham, UK: Flambard.

Wall, G. P., and J. G. Sawkins. 1860. *Report on the Geology of Trinidad, or Part I of the West Indian Survey.* London: Longman, Green, Longman, and Roberts.

Wall, George P. 1866. "The Origin of Bitumen." *Geological Magazine* 3 (23): 236–39.

Watts, Michael. 2001. "Petro-Violence: Community, Extraction, and Political Ecology of a Mythic Commodity." In *Violent Environments*, ed. Nancy Lee Peluso and Michael Watts, 189–212. Ithaca, NY: Cornell University Press.

Watts, Michael. 2004. "Resource Curse? Governmentality, Oil, and Power in the Niger Delta, Nigeria." *Geopolitics* 9 (1): 50–80.

Weber, Max. [1904–5] 1948. *The Protestant Ethic and the Spirit of Capitalism.* Translated by Talcott Parsons. New York: Charles Scribner's Sons.

Weber, Max. [1918] 1946. "Science as a Vocation." In *From Max Weber: Essays in Sociology,* ed. H. H. Gerth and C. Wright Mills, 129–56. New York: Oxford University Press.

Wells, H. G. 1895. *The Time Machine: An Invention.* London: Heinemann.

Wenzel, Jennifer. 2006. "Petro-Magic-Realism: Toward a Political Ecology of Nigerian Literature." *Postcolonial Studies* 9 (4): 449–64.

Weszkalnys, Gisa. 2011. "Cursed Resources, or Articulations of Economic Theory in the Gulf of Guinea." *Economy and Society* 40 (3): 345–72.

Wildavsky, Aaron, and Ellen Tenenbaum. 1981. *The Politics of Mistrust: Estimating American Oil and Gas Resources.* Beverly Hills, CA: Sage.

Williams, Eric. 1944. *Capitalism and Slavery.* Chapel Hill: University of North Carolina Press.

Williams, Eric. 1981. *Forged from the Love of Liberty: Selected Speeches of Dr. Eric Williams.* Trinidad: Longman Caribbean.

Williams, Raymond. 1973. *The Country and the City.* Oxford: Oxford University Press.

Williamson, Harold F., and Arnold R. Daum. 1959. *The American Petroleum Industry: The Age of Illumination, 1859–1899.* Evanston, IL: Northwestern University Press.

Wiltshire, Winston W. 2007. *The Commercial Development of Trinidad Lake Asphalt.* Trinidad and Tobago: Winston W. Wiltshire.

Winer, Lise, ed. 2009. *Dictionary of the English/Creole of Trinidad and Tobago.* Montreal: McGill-Queen's University Press.

Wood, Donald. 1968. *Trinidad in Transition: The Years after Slavery.* London: Oxford University Press.

Yergin, Daniel. 2011. *The Quest: Energy, Security, and the Remaking of the Modern World.* New York: Penguin.

York, Richard. 2012. "Do Alternative Energy Sources Displace Fossil Fuels?" *Nature Climate Change* 2:441–43.

Ziser, Michael. 2011. "Oil Spills." *Proceedings of the Modern Language Association* 126 (2): 321–23.

Žižek, Slavoj. 2010. *Living in End Times.* London: Verso.

Zola, Émile. [1885] 1968. *Germinal.* Paris: Garnier-Flammarion.

Page numbers followed by *f* refer to illustrations.

kerosene, 12, 41–43, 53, 55–56, 56*f*
Keystone XL pipeline, 146, 152
King Coal (Sinclair), 8
Kolbert, Elizabeth, 143
Kublalsingh, Wayne, 112–19, 145–46
Kumarsingh, Kishan, 133–34, 134*f*, 136
Kyoto Protocol, 130

Labban, Mazen, 82
labor and labor power: kerosene labor
  savings rejected by Stollmeyer,
  55–60; laziness, 57, 58; *The Paradise
  within Reach of All Men without
  Labour* (Etzler), 47–48; productivity,
  58; rest–work balance and oil, 58–59;
  slaves and, 30–31; Stollmeyer's
  fecundity with toil, 53; Stollmeyer's
  "paradise without labor," 41–42. *See
  also* somatic power
La Brea: Dome Fire (1928) and decline
  of, 103; gas-fired power plant, 115–17,
  145; as lakeside village, 103; new
  highway route near, 146; as oil town
  and pitch town, 103; positive view of
  hydrocarbons in, 96; smelter debate,
  105–8, 112. *See also* Pitch Lake
La Brea Concerned Citizens United,
  105, 111, 112, 117
Lafargue, Paul, 58
Lashley, Selwyn, 90
Latour, Bruno, 16
Laventille, Port of Spain, 144
laziness, 57, 58
LeMenager, Stephanie, 10
Lenny Sumadh, Ltd., Automotive, Pe-
  troleum, and Industrial Supplies, 108
Li, Tania, 121
Logan, Joshua, 104–5
Lovelace, Earl, 5, 144
Lyell, Charles, 68, 98–99

The Maldives, 137
Manning, Patrick, 111, 116, 131, 133, 135, 137

manure, 156n24
maroons (runaways), 37–38
Marshall Islands, 138
Marx, Karl, 30, 154n8
Marx, Leo, 97
Mason, Arthur, 5, 153n8
Mathews, Andrew, 150
maturation of oil, 75, 76*f*, 92–93
McHalffey, Larry, 83
McKelvey, Vincent, 78–79
McKibben, Bill, 17, 23, 24, 67, 93,
  146–47, 147*f*
McPhee, John, 68
megasse, 54
Merry, Sally Engel, 132
Messerly, Oscar, 99–102, 100*f*
metabolism, 40, 45, 154n8
migration of oil, 73–75, 75*f*, 76*f*
militant anthropology, 4
Miller, Daniel, 20
Ministry of Energy, 83–84, 89–91, 138
Mintz, Sidney, 40
Mitchell, Timothy, 82
Mohammed, Reeza, 110
Monroe, Ethelbert, 104
Montano, Machel, 153n2
Morning Star, 48–50
Mouchot, Augustin, 143–44
Mouhot, Jean-François, 148
Mount Airy plantation, Virginia, 154n8
mules, 35–36
Mumford, Lewis, 54
Munif, Abdelrahman, 10
Myers, Lincoln, 128–30

Naipaul, V. S., 29, 126–27
Nathan, Nathaniel, 101
National Energy Corporation, 110, 115
National Food Crop Farmers Association, 114
natural gas: afforestation for, 145;
  circulation of, 63; converted into
  ammonia and carbon dioxide for

natural gas (*continued*)
  injection, 88; gas audit and reserves,
  83; Heileman on, 129; orange flare
  of, 111; Persad on, 149; Rights Action
  Group and, 115; role of, in Trinidad,
  103; smelter consumption rate,
  161n44
Neanderthals, 86, 159n15
negros cimarrones, 38
New Orleans, 138
Niger delta, 96
Nigeria, 9*f*, 9–10
Nikiforuk, Andrew, 148
Nixon, Rob, 6, 14
Norgaard, Kari, 129
Numa Dessources, George, 52

"Occasional Discourse on the Negro
  Question" (Carlyle), 57
oil: as already cynical category, 60; Darwent's early well, 56; as environmental injustice and structural violence,
  23; "extraction" vs. "production" of,
  65; finders vs. users of, 142; flatness
  of, 43, 60; hidden from view, 6–7;
  labor-saving potential, missed, 57–
  58; leaving something in the ground
  for the future, 135; in literature, 6–9;
  money metaphor for, 10; as new slavery, 148; as salve, 110; somatic power
  from vantage of, 51–53; transition to
  amorality of, 59; work vs. rest and,
  58–59. *See also* asphalt; crude petroleum; hydrocarbons; inevitability,
  myth of; kerosene; petroleum
Oil! (Sinclair), 6, 8
"Oil in the Coil" (Scrunter), 5
oil sands (tar sands), 66, 146, 158n1
Orinoco River and delta, 11, 13, 99,
  123–25
"Orinoquia" region, 123
Orwell, George, 15
Owen, Robert, 46

Packer, Roger, 86
Pantin, Denis, 113, 115
*The Paradise within Reach of All Men
  without Labour* (Etzler), 47–48
pastoral: agrarian vs. small, 113; agricul-
  ture vs. smelting, 114; English rural
  nostalgia vs. Trinidad, 97; La Brea
  and, 103, 118; petro-pastoral, 97–98,
  111, 115
Paul, Tony, 85
peak oil, 66
Percy, Charles, 89–90
Pereira, Vincent, 91
Perrow, Charles, 110
Persad, Krishna, 71–75, 74*f*, 88–94, 135,
  148, 149–50
petroleum, 56, 56*f. See also* oil
*Petroleum Encyclopedia of Trinidad and
  Tobago* (Persad and Persad), 73
*Petroleum Geology and Geochemistry of
  Trinidad and Tobago* (Persad), 73
Petroleum Resources Management
  System (PRMS), 80–82, 81*f*
petromelancholia, 10
petro-pastoral, 97–98, 111, 115
Petrotrin, 89, 106, 109, 136
Phalanx community, 46, 58
pitch. *See* asphalt
Pitch Lake: Agatha Proud as owner of,
  103–5; Asphalt Industry Commission and Wall's surface model vs.
  Messerly's vertical pushing theory,
  99–103; Cazabon's *Asphalt Lake*,
  95*f*, 95–96; historical views of,
  98–99; Logan's play *The Price of
  Progress*, 104–5; Messerly's "chimneys," 99–101, 100*f*; Stollmeyer
  and, 51–53, 58; Wall and Sawkins's
  traverse section of, 69, 70*f*, 99, 100*f*;
  water pollution at, 105–6. *See also*
  La Brea
plantation slavery. *See* slavery
Pointe-a-Pierre Wildfowl Trust, 109

Printed and bound by CPI Group (UK) Ltd, Croydon, CR0 4YY

27/10/2024

14580226-0001